CORONA STAR CATCHERS

The Air Force Aerial Recovery Aircrews of the 6593d Test Squadron (Special), 1958-1972

Edited by Robert D. Mulcahy, Jr.

The 6593d Test Squadron (Special) posing with the Discoverer 14 capsule in 1960./Photo credit: USAF

Center for the Study of
National Reconnaissance

June 2012

Center for the Study of National Reconnaissance

The Center for the Study of National Reconnaissance (CSNR) is an independent National Reconnaissance Office (NRO) research body reporting to the NRO Deputy Director, Business Plans and Operations. Its primary objective is to ensure that the NRO leadership has the analytic framework and historical context to make effective policy and programmatic decisions. The CSNR accomplishes its mission by promoting the study, dialogue, and understanding of the discipline, practice, and history of national reconnaissance. The Center studies the past, analyzes the present, and searches for lessons-learned.

Cover Design: Debbie Bowman designed the cover of this book using photos of a JC-130 aerial recovery aircraft, the launch of a Discoverer/Corona satellite on a Thor/Agena rocket at Vandenberg Air Force Base, and the C-119J loadmasters of the 6593d Test Squadron in the rear of their airplanes during parachute recoveries.

Contents

Foreword v

Preface vii

Introduction ix

Chapters

Chapter 1: Lt Col Harold E. Mitchell 1

Chapter 2: A2C Daniel R. Hill 15

Chapter 3: Capt Robert D. Counts 79

Chapter 4: Capt Donald R. Curtin 105

Chapter 5: SSgt Charles J. Dorigan 127

Chapter 6: SMSgt Richard C. Bell 165

Chapter 7: Maj Gen Donald G. Hard 183

Chapter 8: Col Harlan L. Gurney 199

Appendices

Chronology of the 6593d Test Squadron (Special) 205

1964 Mission Statement of the 6593d Test Squadron (Special) 209

Squadron Commanders of the 6593d Test Squadron (Special) 211

1958 Aircrews of the 6593d Test Squadron (Special) 213

Discoverer/Corona Recovery Pilots of the 6593d 217

Original Assigned Aircraft of the 6593d Test Squadron (Special) 223

Flying Hours of the 6593d Test Squadron (Special) 225

Awards and Honors of the 6593d Test Squadron (Special) 229

Squadron Emblem 231

Documents 233

Glossary 247

Reference Notes 253

Index 257

Foreword

The Air Force personnel who flew in the C-119J and JC-130 aircraft were an essential piece of the Corona story. They were key to the recovery of the film capsules that the Corona photoreconnaissance satellites of the 1960's were returning to earth from orbit. The officers and enlisted crew were the elite few. Their stories not only give us a view inside the operation of Corona capsule recovery, but they also provide insight into how to carry out an efficient and effective Air Force program. These recollections provide an invaluable story during one of the most significant times in the history of national reconnaissance. The Intelligence Community and general public will gain knowledge and an appreciation of those who served.

We are pleased to have accepted Robert Mulcahy's manuscript for editing and publication as a book in the Center for the Study of National Reconnaissance (CSNR) series, *In The Words of Those Who Served*. Robert D. Mulcahy, Jr. is well-versed in aerospace history. He is a civilian Air Force historian for the Space and Missile Systems Center (SMC) at Los Angeles AFB, California, a subordinate unit of Air Force Space Command. Mulcahy has been a historian at SMC since 2000 and has published numerous history-related magazine articles. He earned his B.A. and M.A degrees in history from the California State University at Fullerton.

In this book, Mulcahy delivers a collection of captive narratives from the crew members who were part of this historic time in the history of national reconnaissance. Most of them were unaware of what was in the capsules they recovered, the true mission of the Discoverer program, and Discoverer's relationship with the classified Corona photosatellite reconnaissance program; however, they all understood the importance of their mission to recover capsules from space. The reader will have an opportunity to experience these missions through the perspective of those who served.

I challenge you as you read these recollections to look for lessons in this part of the Corona program—lessons that you can apply to your future challenges. The Corona program tested the limits of technology, stretched the skills of those involved, and overcame disappointments along the way. The perseverance and resourcefulness of everyone involved, from the concept engineers to these air crews who caught "a falling star," demonstrates that the unimagined can become possible and challenges along the way can be overcome.

Robert A. McDonald, Ph.D.
Director, Center for the Study of National Reconnaissance
Business Plans and Operations Directorate
National Reconnaissance Office

Preface

The Discoverer/Corona reconnaissance satellite was a revolutionary breakthrough in intelligence gathering and altered the course of the Cold War. From 1960-1972, the film from these satellites provided vital military intelligence to the United States about the closed interiors of the Soviet Union, Eastern Europe, China, Vietnam, and others. The United States used space reconnaissance to finally determine the truth of the Soviet intercontinental ballistic missile threat, among many other discoveries. With all the information gained from Corona, the United States developed military strategies and national policies to contain the communist threat. The recovery of the Corona space capsules was among the highest priorities of the U.S. Department of Defense and the Intelligence Community.

The 6593d Test Squadron (Special) conducted one of the most important Air Force missions of the Cold War. The aerial recovery of Corona space capsules and various other recovery projects is not widely known. The 6593d was one of the first Air Force organizations to combine air and space as integral parts of its mission. The classified status of the Corona program did not permit the details of Corona or its aerial recovery to be published until the declassification of Corona in 1995.

The objective of this book is to present first-hand accounts of 6593d Test Squadron (Special) aircrew veterans who flew modified C-119J and JC-130 cargo aircraft out of Hawaii when they recovered Corona space capsules and other payloads during the Cold War. The chapters include the perspectives of 6593d officers who piloted or navigated the aerial recovery flights, and enlisted loadmasters who operated the aircraft recovery equipment to snag the parachutes and space capsules and recover them. There are other publications about the Corona program, but this may be the only one that specifically focuses on the recovery aircrews of the 6593d as they tell their story.

The veterans were interviewed (orally or by using e-mail), the transcripts were reviewed by the veterans and edited, and the format was written as a first-person narrative. This book is an important primary source for the Discoverer/Corona space program and provides the experiences and detailed descriptions about how the aerial recovery mission was conducted. The accomplishments of the 6593d are now about 40 to 50 years old, and the experiences of these veterans needed to be recorded for their significance to the Cold War.

The 6593d veterans featured in this book were indispensable in making this publication a reality. They were generous with their time, knowledge, and photographs. Capt Donald Curtin was the first veteran interviewed for the book and helped increase my initial interest in aerial recovery. He also introduced me to some of the other veterans. Also, SSgt Charles Dorigan made very important contributions to this book. He was extremely knowledgeable and answered countless questions about aerial recovery, reviewed the interviews, and introduced me to Dr. Robert McDonald, Director, Center for the Study of National Reconnaissance at the National Reconnaissance Office. Finally, A2C Daniel Hill was also very helpful with information and reviews. My gratitude and thanks to all who provided assistance for this book. The veterans of the 6593d deserve to be recognized and honored for their vital roles during the Cold War.

Robert D. Mulcahy, Jr.
Editor

Introduction

In the late 1950's, the United States earnestly began development of reconnaissance satellites that could capture images of territory controlled by Cold War adversaries. The United States focused early development efforts on satellites that would transmit "read-outs" of film images back to the earth. Because of technological issues, scientists and engineers developing the satellites determined that obtaining imagery from space could be hastened if the satellites' film was returned to earth for development rather than "read-out" on the satellite. Those same scientists and engineers developed return vehicles that would carry the film back to earth. One major challenge remained—the safe retrieval of those film return vehicles. Their safe return depended upon U.S. Air Force crews who would capture the vehicles in midair. In the pages that follow, you will read their words as they recount this remarkable feat that allowed the United States to obtain early imagery from space, and thereby increase U.S. understanding of the intentions of Cold War adversaries.

Reconnaissance was quickly becoming a strategic resource for Cold War combat. In the United States, the President, executive branch officials, military leaders, and members of Congress faced the daunting challenge of understanding the growing Soviet arsenal and appropriately allocating U.S. resources to counter the arsenal. During this period for example, U.S. leaders were first confronted with the perception that the Soviets were outpacing the United States in production of strategic bombers and later with the same perception about Soviet Intercontinental Ballistic Missiles (ICBMs). The Central Intelligence Agency (CIA) developed the U-2 high-altitude reconnaissance plane, which obtained imagery confirming there was not a significant gap between the number of U.S. and Soviet strategic bombers. The Soviet downing of Gary Powers' U-2 in May 1960 and Eisenhower's halt to subsequent U-2 overflights of the Soviet Union, left the question of ICBM imbalances an open one and one that would have to be answered from space.

The CIA in cooperation with the U.S. Air Force was developing the Corona imagery satellite, known then to the public as the scientific experimental Discoverer satellite. The CIA and Air Force experienced thirteen failures before successfully launching and retrieving the Corona return vehicle. With Corona mission 14, the United States would for the first time retrieve film from space by capturing the return vehicle in midair. The Corona program would, in short order, put to rest concerns that the Soviets were outpacing the United States in ICBM production.

The phenomenal achievement of reliable imagery from space depended on the extraordinary skill of the Air Force crews who recovered the film return vehicles. Those crews trace their experience back to earlier efforts to capture descending objects midair. An early imagery reconnaissance program, known as Genetrix, was developed to send cameras over the Soviet Union attached to high altitude balloons. Those balloons were to be captured over the Pacific Ocean once they traversed the Soviet Union. The Air Force crews also retrieved high altitude balloons as they descended after capturing radiologic materials from Soviet nuclear testing. These balloon programs provided an important experience base for the challenge that would follow—capturing a rapidly descending object from space.

The crews spent many hours practicing their retrievals as simulated film recovery vehicles were dropped at higher altitudes for the recovery crews to capture midair. They developed a camaraderie during this time that was essential for making adjustments necessary to perfect the midair retrieval of objects from space. You will find the sense of purpose and dedication of these crew members as you read their words in the pages that follow.

The Center for the Study of National Reconnaissance developed the *In the Words of Those Who Served* series to offer readers more direct access to the reflections of those who served the nation by developing national reconnaissance systems. The oral histories captured in this series are informative in a unique way as you read the words of those responsible for retrieving secrets via space satellites. The absence of analysis and corroboration allows you to witness the unvarnished enthusiasm and perspective of these individuals.

Corona Star Catchers' author, Robert Mulcahy, conducted numerous interviews for this publication. He very ably edited those interviews into a narrative that reveals the challenges associated with midair retrieval of film return vehicles. His work provides an important oral history collection on this essential component of U.S. space imagery programs. Midair retrieval of space imagery would continue until the mid-1980's with the Gambit and Hexagon imagery satellite programs. Mulcahy's *Corona Star Catchers* sheds light on how the crews were developed and sustained in this decades-long effort to catch the nation's imagery from space. These star catchers became instrumental in helping the nation fight and win the Cold War.

James D. Outzen, Ph.D.
Chief, Historical Documentation and Research
Center for the Study of National Reconnaissance

Chapter 1

Lt Col Harold E. Mitchell

Capt Harold Mitchell flying a C-119J in 1959.
/Photo credit: USAF

"I told him to report that Discoverer 14 was safely onboard and we were heading in..."

Lt Col Mitchell (1925-) was interviewed on October 1, 2003. As a captain, Mitchell was a member of the 746th Troop Carrier Squadron from 1954 to 1956 during his assignment as a C-119 aircraft commander for the aerial recovery mission of Genetrix[1] reconnaissance balloons. He flew the Genetrix recovery patrols in Alaska. Capt Mitchell later helped assemble the original members of the 6593d Test Squadron (Special) in 1958. He was assigned as the commander of A Flight for the squadron. Capt Mitchell piloted the first successful aerial recovery of a space capsule when his C-119J recovered the Discoverer 14 capsule on August 19, 1960. He and his aircrew were featured in the national news for the historic recovery. Mitchell was assigned to the 6593d Test Squadron (Special) from 1958 to 1962. Capt Mitchell's C-119J #18037 and the Discoverer 14 capsule are on display at Wright-Patterson Air Force Base (AFB), Ohio, within the National Museum of the United States Air Force

---

Mitchell became the first pilot assigned to the 6593d Test Squadron (Special). All nine initial pilots of the Discoverer program had also flown with the Genetrix program. According to Mitchell, it made sense to utilize their past experience in aerial recovery. Mitchell describes the beginnings of the new squadron and how it was formed.

I left Kodiak, Alaska, on the May 7, 1956. The unit was deactivated. Nine of us went to Pope AFB, North Carolina, as instructor pilots. The nine of us that went to Pope AFB were: Jim McCullough, Larry Shinnick, Ed Mosher, Jack R. Wilson, Tom Hines, Jim Brewton, Warren Schensted, Lynnwood "Lindy" Mason, and myself.[2] Most of us went to different squadrons. We were all instructor pilots, or squadron officers, or flight commanders. I was the initial member of the 6593d. In June of 1958, I had been at Fort Campbell, Kentucky, with a flight of six airplanes working airborne paratrooper drops all week. I got in on Friday night and Ed Mosher, who was my assistant squadron operations (ops) officer, said that I had a phone call from Gen Theodore Kershaw.

---

1   The Genetrix project (Weapon System 119L) that deployed reconnaissance balloons over the Soviet Union also used the code names: Grand Union, Drag Net, Moby Dick and Grayback.
2   These pilots became the original C-119J recovery aircraft commanders for the 6593d.

I went up to see Gen Kershaw. It was on the Friday before the weekend of the July 4th holiday. He asked me what the Drag Net project was so I explained it to him. He told me, "Get one of your airplanes set up to take you to Langley AFB on Sunday." Gen Kershaw gave me my instructions, my walking papers, and said, "Get going." That was it. I don't know why he picked me, although he knew I had been involved with Genetrix. He flew with our squadron quite frequently.

On Sunday, a crew flew me up to Langley AFB, Virginia. I was to meet my contact at Langley, Col Howard Rose who was the Tactical Air Command (TAC) Deputy Director of Operations. We left Langley on Monday morning on a packed Air Force C-47 heading for California. Col Rose was the senior officer, the rest of us were all captains. There was a captain from TAC personnel, a captain from maintenance, a captain from operations, and a captain from supply.

When we got pretty close into Kirtland AFB, New Mexico, Col Rose said, "We have to land at Kirtland. We're having a little bit of airplane trouble." It didn't seem strange to me that night at all, but after I have gone back and thought about all of the things that happened and the people involved… Well, as soon as we landed, a staff car was there, and not just a motor pool staff car, but the base commander's staff car. Brig Gen Osmond J. Ritland, Air Force Ballistic Missile Division (AFBMD) vice commander, was there. They picked Col Rose up and he told us to go to the bachelor officer quarters and get our clothes changed. When Col Rose came back in about an hour, he had transportation requests for us to catch Trans World Airlines out of Albuquerque to Los Angeles. He and Gen Ritland spent quite a long time together.

We got into Los Angeles and the people from AFBMD picked us up. We went from there right into a briefing on June 30th. We met with Maj Gen Bernard Schriever, Gen Osmond Ritland, and staff members who briefed us on the mission concept of using C-119J aircraft for the aerial recovery of nose cones with cameras from orbiting satellites, program code named "Corona." We were all cautioned immediately about the sensitivity of the program and any reference to Corona—none of the briefing information was to leave the confines of that room. As a cover, the Corona program would be called "Discoverer" and oriented to research and development and biological space research using mice, monkeys, and special instrumentation.[3] When we left AFBMD, it was a done deal. TAC would support the Discoverer program. Col Rose and his group returned to Langley AFB to report to TAC, and I reported to Gen Kershaw at Pope AFB.

> *"We were all cautioned immediately about the sensitivity of the program and any reference to Corona…"*

I was, at that time, identified as the TAC project officer for Discoverer. I worked from Pope AFB where I left my job as the squadron ops officer. The next step was to identify the personnel and equipment to outfit this new one-of-a-kind organization (Project Hot Hand).

It is clear that Mitchell wanted quality personnel for the squadron. He traveled to several Air Force bases to request new squad members. Some were former Drag Net personnel, and some were completely new to aerial recovery. Mitchell recalls the process, the places, and the time it took to form the squadron.

When I came back from California, I went to see Gen Kershaw. I just openly said, "We need nine pilots, and we have nine pilots here who are all instructor pilots or squadron ops officers." I identified the nine of us

---

[3] Corona was the top secret code name for the Discoverer program, and only a very select few were aware of this designation until the declassification of the Corona program in 1995.

from the Genetrix balloon recovery project as the Discoverer pilots. Gen Kershaw was a very outspoken gentleman. He said, "By God, you nine pilots can go, but you leave my enlisted people alone." We were the first nine, and all captains, so all of the aircraft commanders came from Pope AFB. We wanted the nine pilots who had already been on the Drag Net project, so it was just a matter of getting back into aerial recovery.

Next I went to the Ninth Air Force Headquarters at Shaw AFB, South Carolina. When I got down there, I met a brigadier general who was assigned by TAC, probably the head of the project. I met with him, and personnel and maintenance officers. We sat down and discussed obtaining the different people. They spread them out through TAC, Sewart AFB, Tennessee, Ardmore AFB, Oklahoma, and other bases. The general agreed to take the nine pilots from Pope AFB. It would not be as big a loss to Gen Kershaw as losing his navigators or his maintenance personnel.

Most of the copilots came from Sewart AFB where they were C-123 pilots. Our navigators were from Military Airlift Command because they wanted navigators with a lot of over-water navigation experience. The navigating equipment on the C-119 was so primitive that it didn't take them long to get acquainted with that.

We were fortunate to have some of the original Drag Net recovery enlisted personnel, the backend crews there, because they could train the new people who were coming in for Discoverer. Some of them were the original winch operators from the Genetrix project, maybe three or four of them. My winch operator was also my aircraft crew chief, TSgt Louis Bannick, who was excellent in both jobs. Algaene Harmon was my chief pole handler. I believe he was also on the Genetrix project. A lot of our recovery crews were brand new, such as A2C Danny Hill, A2C Lester Beale, A1C Bill Gurganious, and young A1C George Donahou from Arkansas. I think a lot of them came from Sewart AFB, and probably some from Ardmore.

I spent the month of July with my fanny tied to the seat of a C-119 traveling between Pope AFB, up to Langley, and back to Pope and Shaw AFB. We worked to get our personnel lined up, which was the biggest problem. It was not an easy task, but one that was accomplished; we did it in thirty days. August 3 was our reporting date for Edwards AFB, California, so you can say from June 30 until August 3, we identified everybody for the 6593d Test Squadron (Special).

Since all nine Discoverer pilots were participants in the Genetrix program, aerial recovery was not a foreign concept. However, recovering a Genetrix gondola was different than recovering a Discoverer capsule. They had to come up with new techniques. Mitchell explains the training, the "how-to" of capsule recovery, and his own methods of accomplishing the mission.

A Genetrix "gondola" (parachute package) weighed about 1,400 pounds. When you tried to recover a 110 or 140-pound package with steel cable on the winches, the cable all came winding back in on you. Contact with the parachute caused a backlash and snapped the cable. In that regard, aerial recovery was like fishing. You get a backlash, and it comes all back in on your reel and that's what it did with the light payloads.

So, we ended up determining new parachute recovery techniques with the technical representatives from Lockheed, who were invaluable. They were very inventive, and used a lot of ingenuity and engineering experience to determine the size of nylon to use as the aircraft parachute recovery line rather than using the steel cable. We also put troughs into the backend of the aircraft, and then interwove the recovery line back and forth in this trough to bring down the G-loads. We could break the G-loads down by tying this nylon 5/8-inch line through the trough. Using the parachute cord broke down the G-loads forced on the capsule when we made contact with the parachute at about 110 knots.

Capt Mitchell with his 6593d aircrew and their C-119J in 1960, left to right: (front) Capt Mitchell, Capt David Torgerson, 1st Lt Robert Counts, SSgt Arthur Hurst, A2C Thierry Franc, (back row) TSgt Louis Bannick, SSgt Algaene Harmon, A2C George Donahou, A2C Lester Beale, and A2C Daniel Hill./Photo credit: USAF

Capt Larry Shinnick was assigned the additional duty as the Recovery Detachment Officer for the 6593d, and they set up a good training program. Discoverer capsule recovery was more of an effort than we had earlier anticipated, because of the change in the recovery equipment.[4] Larry Shinnick had a good knowledge of the recovery gear from the Genetrix project. He was the number four aircraft commander in my flight. Although Larry was one of the recovery pilots, after flying his missions he was running the recovery section. We had to evaluate our equipment, the changes in the parachutes, techniques, and things of that nature. If you have the high experience level of the pilots that we did, it worked out pretty well.

The 6593d aircraft commanders were competitive. When we were at Edwards, not doing the actual missions, Larry Shinnick, Tom Hines, Gene Jones (who was our squadron ops), Jack Wilson, Jim McCullough, and myself all rode in a carpool from Lancaster out to Edwards. If we were out practicing and we missed a parachute recovery, then we had to put a dollar in the pot. Then when we had so many dollars in the pot, we'd stop at a little bar about halfway into Lancaster and blow the pot on beer and shuffleboard, and then came home to our mad wives.

---

4   The recovery equipment, or the recovery rig, consisted of the two 34-foot recovery poles with the attached trapeze-like parachute recovery line and attached hooks that were lowered from the rear of the recovery aircraft to catch the parachute and its attached payload.

Chapter 1 - Lt Col Harold E. Mitchell

We finished our training at Edwards AFB, so we flew to Travis AFB, California, before our flight to Honolulu. It was Saturday night at Travis AFB, and it was raining to beat hell with the winds starting to pick up. We taxied out and Jim Brewton had a problem and then Jack Wilson had radio problems. I had a clean airplane (C-119) so I took off from Travis AFB at about 1:00 a.m. on Sunday, which was on the December 7, 1958. It was nasty weather. We got into Honolulu that evening.

We started training in Hawaii after the first of the year in 1959. The first thing you wanted to do when you located your capsule, was to make a fly-by on the parachute, check its rate of descent, and see the condition of your parachute, if it's steady. Then you did your pre-recovery checklist, opened your beavertail doors, put your poles in the actuators, put them in the slipstream, dropped the poles down, and made an outbound pass. When the poles were all the way down, and you flew the recovery pattern as All American had designed it to be flown, normally the parachute was maybe 30 feet below the airplane when you went across the top of it. When the airplane does that, then the parachute goes into the nylon loop and the parachute was immediately collapsed, torn, and it trailed out behind you.

I had ideas of my own, you might say "techniques." I liked to recover a parachute close up to the belly of the airplane. They didn't like that because you could invert the parachute. I worked it out with my crew. Many times when the parachute went through, it passed close under the belly of the airplane, and went

Capt Mitchell's C-119J about to recover a training parachute./Photo credit: USAF

C-119J loadmasters recovering a training parachute in 1959./Photo credit: USAF

Capt Mitchell's C-119 (#18037) at Hickam AFB in 1960./Photo credit: USAF

over the top of the loop and it wouldn't deflate. It became a drag chute. It took all of my recovery gear and wiped it out. The parachute sucked it on out. To get around that, my winch operator and I got together, and he said, "Why don't I put our poles more in trail. Instead of dropping them down at 45 degrees, drop them down to maybe 30 or 35 degrees." So we did it.

We succeeded in doing two things. A lot of times, if your parachute hit the pole at the right position, it broke the pole off or bent it because of the air in the parachute. We never had a bent pole, because the poles were not down as far. When that parachute hit the pole, it wasn't such a jolt that it snapped the pole, it slid down the pole. I went a year flying practice missions and never missed a parachute. I normally caught them on the first or second pass. My commander didn't like me trailing my poles. Nobody knew I was doing it until they put cameras into the backend of the airplane.

To get our training in, one flight was scheduled to do practice recoveries. They'd go out and pick up two parachutes. A Flight went on Monday. B Flight would take an airplane to 18,000 feet and drop parachutes with simulated payloads on them. Then we recovered the parachutes like we were recovering a regular target.

When it was my flight's time to drop the targets, I'd have them put on maybe three, four, or five extra training parachute targets. I'd climb up to 18,000 feet and have my crew throw a package out. When they said, "The Package is gone," I'd pull my power off, drop some flaps, and they'd start putting the rig out the backend of the airplane. By the time we'd gone out and come back in on the parachute, we were ready to recover it. So, we went ahead and made our recovery, and then put that one away, and moved back up and flew another recovery. We'd practice and maybe get three or four practice missions in, and the other aircraft would get one or two.

You had to make at least one or two recoveries a week with the crew. I think I had 120 parachute recoveries in 1959, something like that. I am just guessing. I would say 200 total practice recoveries. There was nothing really difficult about piloting the airplane to catch a parachute—practice. The Corona parachutes were as solid as the Rock of Gibraltar. They had a nice rate of descent. There was no wiggle in the parachute. It was nothing to recover one.

Flying the C-119 was a little different. I think a lot of it was flying a two-engine airplane 600 miles out over the water. That was the most difficult, because if you were in trouble, you were pushing everything you had to get it home. It wasn't that the old airplane couldn't do it, but we were awfully heavy when we put all of our recovery gear and our crew onboard.

Mitchell's plane was assigned the primary recovery spot for the Discoverer 13 mission. As he points out in his story, an equipment anomaly caused them to miss the capsule. Discoverer 13 hit the water just as it came into the sight of Mitchell's crew. The seasoned pilot knew, however, that his crew's performance was top notch. He recounts the events of that day.

On each Discoverer recovery mission, the flights rotated their aircraft positions in the recovery zones. For Discoverer 13 my flight (A Flight) was assigned the north recovery zone, which was the primary recovery area. It was 200 miles long, and you had to assign the flight in order. Then on the next mission, B Flight moved up to the most desirable recovery zone, and A Flight then took the 400-mile area. For Discoverer 13 our flight aircrews had the optimum recovery area for that mission. I took the number one slot, which was the optimum slot. Then on the next mission, I would take the last slot, and every aircrew moved up, so that everybody had an equal chance to be in the primary recovery zone.

The Discoverer 13 recovery area was kind of northwest of Honolulu, and my crew was scheduled to have the number one slot. When the Discoverer 13 capsule started to descend and its homer came on, my navigator picked up the signal, but the receiver was so saturated that we couldn't tell where Discoverer 13 was. We were flying a circle below it, but one of the control aircraft, an RC-121 from the 552d Airborne Early Warning and Control Wing, said he had a target for us and put us on a heading of 285 degrees. So, we flew off at 285 degrees.

On recovery missions I'd take my airplane up to 18,000 feet. When the operation started, I'd start a slow descent, leaving my power up. By the time we had traveled a few miles, we were near the redlined airspeed on the airplane. Our redline was 285 knots and I could be moving at 270. Our C-119, she would get awful stiff, but the old bird would get up there.

We started our run. We had been on the recommended flight path for about five minutes when Bob Counts, my navigator, asked me to do a 360-degree turn, which I did. We had no more than made a 180 of the 360-degree turn, when Bob said, "The target is behind us." The RC-121 came on the air and agreed with Bob. The RC-121 had given us a reciprocal heading when he sent us off on the run at 285 degrees. We got back just as Discoverer 13 hit the water. Larry Shinnick was flying over the water tracking it.

Of course, there was disappointment on the part of my crew after Discoverer 13, but I was more concerned about the saturated signal that started the chain of errors. I told the people back at the base that the receiver from General Electric was saturated, because the capsule was coming down directly over our heads and you couldn't get a direction from it. They wouldn't believe us.

The following week we flew our aircraft #037 nearly everyday simulating the problem we experienced with the beacon the week before. They sent a B-47 out from Edwards AFB and he climbed up to altitude with a beacon on it, and we'd go out and track the beacon to check our equipment. We had the tech rep from General Electric in Philadelphia with us, and we couldn't find anything wrong with our equipment and finally determined that it was the nature of the antenna. For the time being, we had to work around it. So, they said the peculiarity of the system was something that we were just going to have to live with. So much for high tech design!

Well, it felt like we were blamed for not recovering Discoverer 13, to begin with. In my opinion, my crew's performance was the best on the flight line. When I left the 6593d, everybody wanted my winch operator and my navigator. I had a lot of good navigators in my various crews, but Bob Counts sits at the top of my list as the best of the bunch, an outstanding navigator and an excellent young man. When that airplane went in for inspection, it didn't go just with the maintenance people. The pole handlers and everybody else went with it. It got a thorough washing down inside and out. You could eat off the floors in it, and it was always polished to a high gleam, there wasn't any oil on it. If oil showed, and the Wright engine was known for being oily, it was wiped clean. The airplane was immaculate. It was just a great crew, and I never heard any of them ever have a cross word, or a disagreement, or anything else. It worked out alright.

> *"In my opinion, my crew's performance was the best on the flight line...It was just a great crew, and I never heard any of them ever have a cross word, or a disagreement, or anything else. It worked out alright."*

Chapter 1 - Lt Col Harold E. Mitchell

The day before Discoverer 14's reentry, Mitchell's C-119 had mechanical problems. However, by the morning, he was notified the plane was ready to fly. This string of good fortune continued for the crew of Pelican 9 (#037). With pride, Mitchell recaps the events that occurred during the historic mission of August 19, 1960.

The antenna turned out to be the least of our problems. Discoverer 14 was launched on August 18th and on that same day TSgt Bannick, my maintenance chief and winch operator, found an intake leak on the number-four cylinder in the left engine of #037. That meant a cylinder change and none were available at Hickam AFB, Hawaii. Mission planning for the Discoverer 14 reentry went on the next day.

On the morning of August 19th, we reported to ops at 6:00 a.m. for a 7:00 a.m. briefing. Bannick met me in his flying suit with a big smile on his face, so I knew he had #037 ready to fly. He always did. The briefing was the usual: aircraft position, departure times, codebooks, and a padlock for the nose cone shipping canister. Since A Flight had the primary recovery area on Discoverer 13, we had the secondary area for Discoverer 14, and my crew position was Pelican 9. At the briefing, we were told the Agena vehicle was in an abnormal attitude and was using control gas at an excessive rate. This meant the reentry vehicle would be affected if there wasn't sufficient gas to position it correctly for a reentry. As I was to discover later, this became a boon for us on Pelican 9.

We took off from Hickam at 9:00 a.m. with A Flight. Each plane dropped off to assume their orbiting position in the recovery pattern. As Pelican 9, we proceeded 300 miles southwest of Hickam. When we arrived on station, my copilot Capt Richmond "Rick" Apaka and I went over the operating procedures, the code numbers for reporting the different postures of the missions, and how I intended to fly the

Discoverer 14 capsule being recovered./Photo credit: USAF

recovery. This was only Rick's second operational mission. He had been with me on Discoverer 13. After the recovery crew checked their equipment, we climbed 16,000 feet and listened to the mission progress on the command post frequency. At 12:46 local time, the command post announced that the ejection of the reentry vehicle over Kodiak, Alaska, had occurred.

At 12:53, Counts advised that he had a beacon signal on a heading of 255 degrees. Rolling out on a heading of 255, Pelican 9 started picking up speed in a gradual descent and the controls were feeling pretty stiff. I checked the airspeed at 275 knots, 15 knots below redline. Counts asked for a 360-degree turn to check the signal, and at its completion he confirmed 255 degrees for intercept. Within a couple of minutes, dead ahead and 4,000 feet above us was the orange and silver chute with a gold capsule the shape and size of a kettle drum gleaming in the sun. Rick called the command post with the code number for a visual sighting, but they were too busy vectoring another plane to a suspected target to hear our transmission. From then on, it was just us. We proceeded with the job before us without further command post notification.

When the aircraft was slowed to 120 knots, I had the beavertail doors opened and the recovery rig extended and lowered into position. When all was ready, Bannick read off the recovery checklist and added one admonition, "Captain, good luck and for gosh sake's, don't invert it!"

As the parachute came through our altitude, I rolled in our first recovery approach. Nearing the aircraft, the chute loomed in the windshield and passed just below the belly, but I didn't feel the slight tug of a contact. SSgt Harmon, chief pole operator, said, "Six inches off the right pole." Flying another pattern and approach for a second recovery attempt, I was two feet too high of the parachute. Below us was a deck of stratus clouds with tops of about 7,500 feet. This could be the last pass, so I rolled in on my approach 800 yards from the target. As I rolled level, the chute was bobbing and weaving a little and moving left, so I edged the plane bit-by-bit as the chute flashed down the belly of the fuselage, then I felt that slight tug. Harmon came in on the intercom again, "Good hit captain! We've got her in tow!"

> *"As I rolled level, the chute was bobbing and weaving a little and moving left, so I edged the plane bit-by-bit as the chute flashed down the belly of the fuselage, then I felt that slight tug. Harmon came in on the intercom again, 'Good hit captain! We've got her in tow!'"*

I had Rick report a successful recovery back to the command post, but we were told to stay off the air and not to interfere with the recovery attempt in progress. The northern aircraft were still chasing a false target. I told Rick, "OK." I sent each crew member on the flight deck below to see our catch before the capsule was placed in a gray canister and securely padlocked for its trip back to the States.

When Rick came back to the flight deck, I went down to congratulate the crew on an outstanding job, and thank them for all their patience and the long, hard hours of flying. TSgt Bannick reached into the front of his flying suit and pulled out a torn piece of orange nylon parachute and handing it to me said, "For you captain. They will never miss it." I still have it, or most of it. When I looked at the top of the gold capsule in the gray canister, the top was covered with soot from the retrorocket, and you could see etching under the soot with the names of those who packed the recovery parachute. When I returned to the flight deck, Rick said the command post was requesting our status. I told him to report that Discoverer 14 was safely onboard and we were heading in, giving them our estimated time of arrival.

Capt Mitchell congratulating his crew for the recovery of Discoverer 14 as they flew back to Hawaii. Seen in the photo are Capt Mitchell, A2C Dan Hill, TSgt Louis Bannick and A1C George Donahou. (Photo by SSgt Wendell King) /Photo credit: USAF

Gen O'Donnell awarding Capt Mitchell the Distinguished Flying Cross. Photographed members of the Pelican 9 crew, left to right: Capt Mitchell (aircraft commander), Capt Richmond Apaka (copilot), 1st Lt Robert Counts (navigator), SSgt Arthur Hurst (flight engineer), TSgt Louis Bannick (winch operator), and SSgt Algaene Harmon (loadmaster)./Photo credit: USAF

It was a jubilant crew that brought the aircraft #037 home that evening. As we flew homeward, we were joined, as always after each mission, by the other three aircrews of A Flight who formed into a tight diamond formation back to Honolulu. As we neared the field, they dropped into trail behind Pelican 9.

When we taxied in, there was a crowd of family, squadron members, and press to greet us. Gen Emmett "Rosy" O'Donnell, the Pacific Air Forces commander, said he had called Gen Thomas White, Air Force Chief of Staff to report the recovery, and he said, "I don't know these men, but give them medals!" I was presented the Distinguished Flying Cross and each crewmember received the Air Medal on the spot.

The excessive use of gas by Discoverer 14 to maintain attitude was a boon to Pelican 9. I guess the lack of sufficient gas to attain the correct reentry attitude made the capsule overshoot the primary reentry area and come down 30 miles from Pelican 9. It was 600 miles long of the intended reentry. We were quite surprised that our aircraft was actually the one in position to recover Discoverer 14.

After #037's successful recovery of Discoverer 14, Mitchell and crew became instant celebrities. Not only did they receive medals immediately following their return, they were whisked away on a promotional tour. Mitchell appeared in his hometown parade as well as a being a guest on the Ed Sullivan show. Lt Col Mitchell concludes with reflections on the invaluable role of Discoverer 14 and the crew that recovered the capsule.

After we made the recovery, and they had a press conference, the 6593d had a beer bust. They had a keg of beer there in the hangar. It was a pretty long seventy-two hours, really. We had been promised that whoever made the first recovery, the pilot, and whoever he wanted to take with him, were invited to come back to the States and go on the Dave Garaway Show. In a laughing way, I told my wife the night before, "You might as well pack my bag, because I'm going to be heading for the States tomorrow." We lived in Kailua and when she came to the base that afternoon, she had my bag packed.

After the keg of beer, my navigator, 1st Lt Bob Counts, winch operator/maintenance chief, TSgt Louis Bannick, and I shaved, showered and put on our class A uniforms. We went over to the Honolulu International Airport and left at 11:00 on United Airlines. I think United gave us a free ride back to Los Angeles.

We got into Los Angles about 7:00 a.m., and Gen Ritland met us with the launch officer Capt Roy Lefstad from Vandenberg AFB, California, who had launched Discoverer 14. He met us at the airplane and walked us into a terminal where they had a press conference. Then we went out to Air Force Ballistic Missile Division Headquarters and went over our program. After the meeting, we met an escort from the public relations office who then put us in a motel. We tried to get some sleep because we hadn't slept since Friday at 4:00 a.m. We didn't sleep on the airplane that's for sure. Then we caught a flight that night back to New York City.

We got into New York City at about 7:30 a.m. on Sunday, and we went directly from there to tape the Dave Garaway Show. Bob, Bannick and I were on the Dave Garaway Show. I don't know how many "kiddy" programs in New York we were on.

The Discoverer 14 recovery on August 19, 1960, Bob White's X-15 altitude record on August 12th, and Joe Kittinger's 102,000-foot parachute jump on August 16th all came in the same week. I went back for the Ed Sullivan Show with Joe Kittinger. We were on the show the Sunday night before Labor Day. At the last minute, after Oscar Hammerstein died, they changed the whole program and had the 1960 Ice Capades. They were going to show Kittinger's parachute jump, the Discoverer 14 recovery, as well as Bob White's

X-15 flight. I was televised with Kittinger since Bob White couldn't come. They also had the president of the American Federation of Labor and Congress of Industrial Organization, George Meany and his wife there.

I went to Air Research and Development Command (ARDC) Headquarters on Friday with Counts and Bannick. We were there all week. We eventually joined my entire crew in Washington, D.C. on Tuesday. White, Kittinger, and I were tasked to brief Gen Schriever's staff on each of our projects, but not at the same time. Also Gen Schriever initiated all of us into the Aerospace Primus Club: my crew, White, and Kittinger. On Friday evening, they had a cocktail hour and dinner for all of us there at Andrews AFB, Maryland. At least I knew, and I imagine Bob knew, but Lou didn't know the ramifications of Discoverer 14.

Bob White, Joe Kittinger and I did quite a lot of traveling. We were all three guests at the Air Force Association's Convention in San Francisco, and then Air Force Systems Command had a large contractors' convention and display in Las Vegas. It had the first showing of the Minuteman missile.

I was born in Bloomington, Illinois, and I was raised in my little hometown of Greenfield, Illinois, which is 120 miles south of Bloomington. Congressman Leslie C. Arends was an old friend of the family. My father went to college at Wesleyan University where he met Leslie. When the mayor of Bloomington found out about the recovery and that I was from Bloomington, he went to Leslie Arends and asked if he could get me and my airplane to Bloomington for a celebration. Congressman Arends agreed and made a request to the Air Force.

I flew a C-119 from Edwards AFB back to Bloomington for a weekend that included a parade and celebration with a formal banquet. They brought the capsule in on a Southern Illinois DC-3. Col Lee Battle

The Discoverer 14 capsule being displayed in front of a banquet table at Bloomington./Photo credit: USAF

from AFBMD came out as a guest speaker. Col Battle gave as much of a briefing on the program as he could, and then he introduced me. We left the next day and flew back to Edwards, and then back to Honolulu.

In Hawaii, my crew and I rode in the Christmas parade. We went out in a boat to where Jarv Adams flew over with a C-119 and dumped a dummy Santa Claus out of the backend. It came floating down on a parachute. When we recovered the dummy, they took us back to the wharf and off-loaded us. We led the parade in convertibles for Christmas.

Our crew stayed in the Corona program through the next year as we transitioned into the C-130 aircraft and continued with the Discoverer series. I was promoted to major in 1961, and I left the 6593d in April of 1962. Yes, the Corona program was an experience in catching "falling stars" for the crews of the 6593d Test Squadron.

I didn't know there was film onboard of Discoverer 14 until I came back to the United States and I was at ARDC Headquarters. The Discoverer 14 film covered 1,650,000 square miles of the Soviet Union, more coverage than all twenty-four U-2 overflights combined, and much of the area had never been reached by the U-2. Corona contributed invaluable information over the years, not just during the Cold War and for military needs, but also for agriculture, mining, conservation and many other uses. It advanced from camera filming to direct readout—a long way from the 100-pound gold capsule resting at Wright-Patterson AFB along with #037.

Chapter 2

A2C Daniel R. Hill

A2C Daniel R. Hill in 1959./Photo provided by Daniel Hill

"Being advised that our role in these very dangerous missions would be to catch a falling space capsule was a total shock to us! Doubt! Disbelief! During those days most folks knew space as something out of a science fiction book or movie."

A2C Hill (1939-) was interviewed using e-mail from June 14 to November 14, 2003. Hill was a teenaged enlisted loadmaster when he became one of the original members of the 6593d Test Squadron (Special) in 1958. He was assigned to Capt Harold Mitchell's C-119J aircrew, and helped recover the Discoverer 14 capsule during the first aerial recovery. Hill accumulated over 1,000 hours of C-119J flight time for the 6593d, and had about 250 aerial recovery flights for training and actual missions. A2C Daniel Hill was assigned to the 6593d from 1958 to 1960.

A2C Daniel Hill was an experienced aircraft loadmaster. He was familiar with flying in the rear of an airplane with the cargo door open. The difference was *loading* into, rather than *unloading* out of, the open door.

I was assigned to the 2d Aerial Port Squadron of Tactical Air Command (TAC) at Sewart AFB as a loadmaster, mainly on C-123 Fairchild aircraft, since July 1957. As tactical support, we flew many missions relocating squadrons and supplies throughout the Air Force. At least half of our time was spent training with the Army 82d Airborne Division at Fort Bragg, North Carolina, and the 101st Airborne Division at Fort Campbell flying in and out of dirt strips with Army personnel and equipment. I know for sure that there were four of us that transferred from the 2d Aerial Port Squadron to the 6593d. Beyond that, I'm not sure. I can't really say that any of the older Genetrix loadmasters were from the 2d Aerial Port Squadron. For some reason, I can only remember the older ones from our time at Edwards AFB.

The few items that I can think of that may have been factors in my being selected for the recovery program are the following. I feel that my loadmaster duties were above average, I was also skilled in working at the rear of the aircraft with the door open while dropping troops and cargo. I volunteered for as many flights as possible. I was then transferred to the 6593d at Edwards AFB. I arrived in September 1958 as an airman second class.

I was briefed that our new 6593d Test Squadron was being developed to train and test equipment to make successful aerial recoveries of various scientific equipment. First, let me say that catching anything with an airplane seemed mind-boggling. When I was introduced to the fact that we would be making midair recoveries of scientific data with lines, poles, hooks and parachutes, who would think that it was possible? I was not at all familiar with the Genetrix project. So the whole concept was doubtful to a young airman eighteen years of age. From a kid out of the fruit orchards of Pennsylvania, I crammed in a lot of information and fascinating things to become a loadmaster, where at times, we also dropped troops and cargo.

Up until this point I was trained to assist in dropping things out of the rear of a flying aircraft, and now I was expected to assist catching something out of the blue and bring it in the rear of an aircraft. The whole concept was different than anything one could imagine. Then shortly into the program being advised that our role in these very dangerous missions would be to catch a falling space capsule was a total shock to us. Doubt! Disbelief! During those days most folks knew space as something out of a science fiction book or movie. Was it possible that these people knew what they were talking about? Was it real that we were now part of this storybook adventure?

Upon arrival at Edwards AFB, in late August and early September of 1958, we were introduced to the newly formed 6593d Test Squadron (Special) Air Research and Development Command (ARDC) by a series of briefings involving the introduction of personnel, aircraft and recovery equipment. All the aircraft commanders, most of the tow reel specialists/winch operators, and a few of the loadmasters had some prior experience of similar recovery operations while on the Genetrix project. Approximately eight to ten of the loadmasters had previous air recovery experience. The aircrews were assigned to a particular aircraft and remained with it throughout our research and development of recovery operations at the Edwards' Flight Test Center.

With all their previous moving around, the C-119s of the 6593d did not initially meet the required operating standards. Hill praises the 6593d aircraft maintenance section for getting the planes into flying condition. These maintenance teams kept the aircraft flying and their flight crews alive.

Records indicated that the C-119s were in "poor" overall condition upon their arrival to the 6593d Test Squadron (Special) at Edwards AFB. This was mainly because the aircraft had not been in any one location long enough in order for them to be put in good operating condition. The C-119s had previously, in 1955, been sent back to Fairchild Aircraft in Hagerstown, Maryland, and then modified for the Genetrix balloon air recovery operations. This was the same facility where they were originally manufactured and tested. In May of 1956, they were returned to the Ogden Air Material Area at Hill AFB, Utah, and stored for approximately nine months. Eventually, they were delivered to Hayes Aircraft in Birmingham for inspection, repair, removal of the recovery equipment, and were then returned to a troop carrier configuration. In late 1957, they were assigned to several reserve facilities. Shortly after that, they were all taken to Fairchild Aircraft at St. Augustine, Florida, for additional modification and the reinstallation of the recovery equipment. As each aircraft was completed, it was then flown to the 6593d Test Squadron (Special) at Edwards AFB, from August 21 through September 25, 1958.

During the same time period that the aircraft were arriving to the 6593d, so were personnel with various skills, and among them were the highly trained airmen that came to form the new maintenance section of our test squadron. These airmen had long-time, previous experience working on the C-119 aircraft. Immediately, these special skilled airmen were divided into flight line teams consisting of a crew chief that was responsible for several aircraft, a mechanic, and a flight mechanic (chief engineer) who were assigned to a particular aircraft. Each aircraft was then thoroughly inspected, repaired and flight line tested by its assigned competitive maintenance team. The chief engineer also flew with us as an aircrew member at all times. These maintenance teams were very serious about what they did and through their competitive nature, perfected our aircraft and maintained them in topnotch condition. Now, we not only had airplanes with recovery gear installed, but ones that could be dependable for many hours of flight time recovery training. It was very obvious to us flight crews, that these dedicated, competitive maintenance teams were keeping us alive and flying.

The 6593d had aircrew personnel who were veterans of the Genetrix project. The experienced squadron members taught the others how to perform aerial recovery. Hill explains, in detail, about the training, the techniques, and everything in between.

We were trained by some experienced personnel who had prior knowledge of the midair recovery techniques from the Genetrix project. Because of the experience of the Genetrix loadmasters, winch operators and aircraft commanders, our training was much easier for the balance of us to quickly understand. We also had briefings and classroom training on recovery operations. The recovery gear was designed and implemented by the All American Engineering Company of Wilmington, Delaware, and with us for much of the training and certification was Harry Conway, a technical representative for this company. Harry was a dedicated, daring individual. He certainly knew and trusted the All American product. Together we helped develop new techniques concerning our midair recovery systems.

> *"Because of the experience of the Genetrix loadmasters, winch operators and aircraft commanders, our training was much easier for the balance of us to quickly understand."*

Of course, everything we did was to be kept under our hats. At first, we could not discuss or write to anyone concerning what we were doing. Day after day we took to the air to practice. We had ten C-119J Fairchild beavertail aircraft, with four assigned to Flight A and six to Flight B. Our crew was the lead aircraft for Flight A because Capt Harold E. Mitchell, our aircraft commander, was also the commander for Flight A. Our practice missions were mostly over the Mojave Desert, away from people and places, to avoid contact with anyone or any activity outside of our squadron.

If a particular crew could not be in the air due to a mechanical problem with their aircraft, they were responsible for the ground recovery of practice equipment. They would drive out into the desert area in several weapons carrier vehicles and be ready below our air recovery operations. Should any air recoveries fail, we would visually track the payload and parachute, then drive to the landing impact area and recover the equipment, so it could be reconditioned for reuse in future practice air recoveries. During the ground recovery of the equipment, in addition to the ground vehicles, we also used ground-to-air radios and binoculars to aid in our search-and-locate duties. Believe me, in the early days we missed a lot of air recoveries for various reasons.

We used several different items for our practice aerial recovery payloads in an effort to determine which was the most effective to achieve the desired weight that was required to simulate what would someday, in the near future, be caught on a real live aerial recovery operation. A few of the experimental weights consisted of a short precut piece of railroad track, auto radiators filled with some concrete, and square or rectangular concrete-poured blocks. Of these choices, the rectangular concrete blocks were preferred

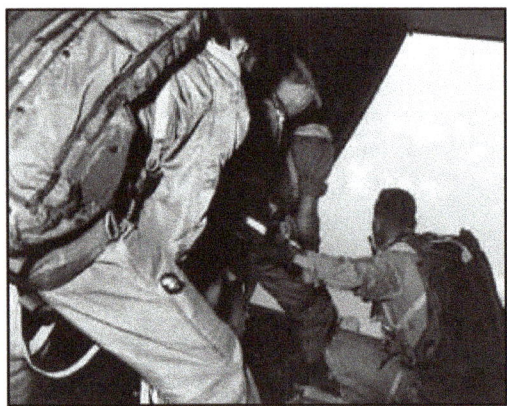 

C-119J loadmasters recovering parachutes suspending concrete blocks during their training./Photo credit: USAF

because they had the best descending parachute speed and the best stable tow of a payload behind the aircraft while being winched toward the rear ramp opening.

Parachute failures were numerous, which resulted in several design changes and modifications. Research found that when special nylon webbing was sewn onto the parachutes in various locations, it reinforced the chute to help avoid the four-pronged bronze hooks from tearing completely through the silk and risers of the chutes. Other design and modifications of the parachutes had to be made to assure a favorable descending speed of the chute and payload at the time of hook contact at recovery. Some changes would help the stability of the chute and payload but the descent-rate speed would then be too fast. After numerous designs, modifications and practice payload failures, a good breakthrough was achieved that provided favorable stability, and the descent-rate of the parachutes was reduced from 1,900 to 1,600 feet per minute.

At this time, we were finding that airspeed and the impact of the recovery gear to the parachute was too much of a shock load (G-force), thus we were having too many recovery failures of this nature. We were using half-inch diameter nylon rope on our recovery winch, and sometimes it would snap in two at contact, and we would lose the chute and the payload, leaving it to plummet to the ground. The rope was wound onto the winch assembly, then exited to the rear of the aircraft, and then it was connected to the balance of our recovery gear that consisted of an additional half-inch diameter nylon rope and three bronze hooks. Two hooks were attached to two long poles, the other was in trail centered between the poles. The nylon rope would stretch at recovery impact, and much of the time it would not snap in two, but this was not a foolproof method because we were losing too many payloads, so more research and development had to be planned.

An aluminum metal trough was designed. The aircraft floor was modified and the trough was recessed into our recovery area floor between the recovery winch and the rear of the aircraft. Then small lengths of nylon cord (100-pound test) were tied around the half-inch nylon rope. The rope was then zigzagged back and forth across the inside width of the trough and the many, many nylon strings were attached to aluminum pegs that were part of the trough assembly. What this procedure accomplished was to gradually absorb the shock through each snapping of the nylon cords, and not a violent sudden impact that severed the half-inch nylon rope. This technique proved successful. At the start of a midair recovery, as the bronze hook or hooks made contact with the descending reinforced parachute, a procedure began

C-119J winch operator TSgt James Cross of the 6593d beside his parachute recovery winch wound with nylon recovery rope in 1960./Photo credit: USAF

 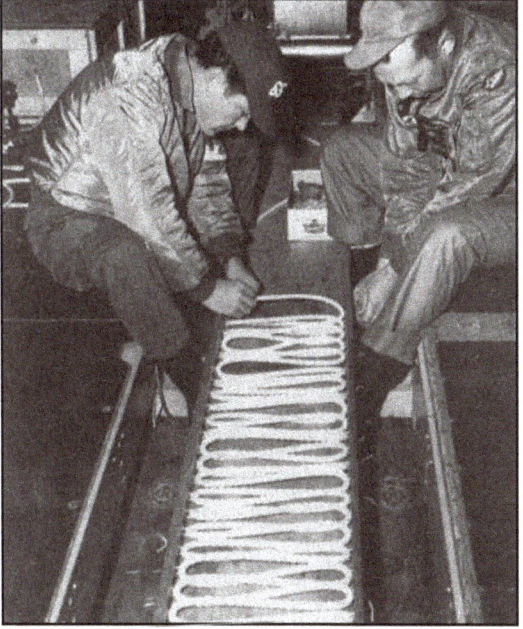
Left: A C-119J winch and trough with the zigzagged recovery rope in 1958. Right: Loadmasters A1C Billy Gurganious (left) and A1C Walter Johnson tying a recovery rope inside a C-119J trough. /Photo credit: USAF

by the snapping of each cord very rapidly until all the zigzagged rope was now pulled out the rear of the aircraft, along with approximately three-fourths of the rope that was spooled onto the recovery winch, thus the payload was in tow waiting to be winched to the rear of the aircraft and pulled on board. What this successful technique provided was, as each snapping of cord took place, the initial shock load was spread over a greater time interval, thereby reducing the peak G-forces. The noise of the cords snapping at impact reminded me of the loud rapid sound of a machine gun firing.

> *"The noise of the cords snapping at impact reminded me of the loud rapid sound of a machine gun firing."*

When time permitted between practice flights, ground recoveries and briefings, we would take our turn at packing the parachutes that we used to make airdrops for practice aerial recoveries.

Both Flights A and B would go out on practice missions, but not necessarily at the same time. Within our own flight, each aircrew would alternate to be the drop cargo aircraft, meaning that we would fly at a much higher altitude than the others. We would open the rear ramp and drop the training payload and parachute while the other aircrews below us would search out and attempt to make aerial recoveries. Should the other aircrews miss the intended recovery, we would perform a partial rapid rigging of our recovery gear immediately after the drop, and then fly an altitude descent as fast as possible to assume the position of backup recovery crew. By rotating these positions time and time again, we accumulated hours of practice for all of the crew requirements and each day we became more efficient, or so we thought. Some days our efficiency bubble would burst when one of our aircrews returned back to Edwards with one of the bronze hooks pronged into the horizontal stabilizer at the tail of the aircraft, or an aircraft returned to base on one engine because the other engine had to be shut down with a large portion of a parachute lodged in it and wrapped within the four-bladed propeller.

Some very risky things happened when flying these missions. Most people would not believe, or certainly would not want to be involved with, some of the hair-raising situations we would find ourselves in. These particular training missions were, or could be, very dangerous to all of the aircrews. The aircraft commanders had to have a keen eye, and depth perception was a must to fly these Boxcar aircraft at the required speed, while in a descending altitude and lining up for the recovery contact on a parachute and payload that was dropping fast at 1,600 feet per minute (27 feet per second). At the rear end of the aircraft were us loadmasters with the recovery gear in trail. We waited for the visual sighting of the parachute canopy below us making a quick contact with one or more of the hooks and listening for the impact noise and payout of line from the G-force reduction trough and the tow winch.

In earlier recovery operations, it was not uncommon for the payload to swirl and swing from side-to-side while in tow behind the aircraft when it was winched near the rear of aircraft. We loadmasters would have to reach out of the rear opening and physically pull the payload inside the aircraft. We always wore a personal parachute when working at the rear of the aircraft and with open skies below us. When recoveries were to be performed over open water, it was mandatory that we also wore a "Mae West" inflatable life jacket under our parachute. Also for personal safety, we were connected with a flat nylon webbing from a D-ring on our parachute to a metal cable that was installed on both sides of the fuselage and spanned our recovery working area. This hopefully would prevent us from being pitched or falling out of the rear of the aircraft.

The C-119J loadmaster in the center is signaling the success of a snagged parachute about to be recovered in 1959. Left to right: SSgt Fred Stebbins, SSgt Thomas Mills, and A3C Hector Santana.
/Photo credit: USAF

   The loadmasters were issued knives that we strapped to our boots, put in a flight suit pocket, or carried at the waist attached by an adjustment strap on our flight suit. We carried the knives for emergency reasons in case we got tangled in parachute risers and had to cut ourselves or another airman free. If a loadmaster got tangled in these lines or any part of the parachute, and by chance the chute inverted near the tail of the aircraft, the airman (if still entangled) would have been pulled out of the plane very quickly if the inversion snapped the half-inch nylon recovery rope, dread the thought. These safety measures fortunately worked because not one of the four-man team of loadmasters on any of our squadron flights was ever lost out the back of an open aircraft.

   During the first phase of the open ramp recovery gear setup, it was the responsibility of the loadmasters to place the 36-foot long booms into the two hydraulic actuators, one on each side of the aircraft. These poles were manually handled to first extend through an opening in each actuator approximately 3 feet outside of the aircraft. We loadmasters (one on each side of aircraft) would sit and straddle the top of the actuators. At this time, we would place one leg out the rear of the aircraft and rest a foot onto an extension guide attachment of the open beavertail ramp while stabilizing our bodies to allow us to stretch and lean out of the rear opening. At this point, one of the four-pronged bronze recovery hooks that were spliced onto the outer end of our half-inch nylon line could be attached to the end of each boom with a strong masking tape that was wound around the shank of the hook and approximately the last 6 inches of the boom. Once the hook was relatively secure, we then clipped the half-inch line to the boom above the hook shank by means of a metal clip pin secured into a prefabricated hole in the boom. The boom was

C-119 loadmaster A2C Frank James of the 6593d wearing a parachute while strapped to the fuselage for safety in 1959. /Photo credit: USAF

Loadmasters watching a parachute recovery from the rear of a C-119J./Photo credit: USAF

then pushed farther through the actuator and more to the outside of the aircraft where we could then attach another secure clip pin approximately 8 to 10 feet further up the length of the boom.

Sometimes when at higher altitudes it was very cold up there, leaning out the rear of the aircraft, and our masking tape would become brittle during the wrapping installation and it would quickly tear. When this condition persisted after several speedy attempts, we would try to secure the hook-to-boom by using new tape from a new roll. It was found that if we could and would relax more, under these sometimes aggravating processes of wrapping, it was much to our advantage and assured a good tight wrap that eventually secured the hook to the boom. The brittle condition mainly existed during the wrapping process only.

With the booms out only through the first phase of the recovery setup, most of the recovery gear still remained inside the aircraft. This would allow the pilots to fly and descend at the highest rate of speed to reach our recovery area, and to allow more time to locate and pinpoint and then throttle down the airspeed. Once the aircraft was down to an acceptable airspeed, we would place the balance of the recovery gear out the rear of the aircraft. The balance of the recovery gear consisted of two more bronze hooks, one that would float to the high center, and the other would trail at the low center between the stationary hooks attached to the end of each boom. By now an additional hook had been developed into our recovery technique for a total of four possible contact points. The floating hooks were attached

Recovery lines and hooks trailing below a C-119J during a 1959 training recovery./Photo credit: USAF

to the same nylon rope harness that was attached to the booms. The two floating hooks trailed dead center behind the aircraft with approximately a 6 to 8-foot vertical spread. The booms were then pushed completely into the hydraulic actuators. This left our recovery gear trailing almost straight out the rear of the aircraft, but at a slight downward angle.

Upon sighting the descending parachute and payload, the aircraft commander made a series of turning banks, adjusted the altitude and airspeed, and maintained a constant descent to assure that the aircraft was aligned with the recovery package. The winch operator checked and made any final adjustments to the winch drum assembly, double-checked the proper drum braking settings, and also activated a firing mechanism that could be triggered to sever the recovery line in case of a severe emergency. The winch operator performed one of the most important functions of the aerial recovery technique. Without knowing the precise time to immediately engage and determine the proper pressure on the drum reel braking system, the impact and rapid payout of the recovery package would be a total failure by either losing all of the line, or a part of the line, due to sudden G-force line severing.

> *"The winch operator performed one of the most important functions of the aerial recovery technique."*

Once the aircraft commander was satisfied with maintaining a descending alignment, he performed a 180-degree descending turn to come back toward the package for a final alignment prior to recovery contact. Before the final alignment turn, we placed the hydraulic actuators, which contained the two booms and the balance of our recovery gear, downward in an approximately 30 to 45-degree angle, based on the aircraft commander's preference and request. Our trapeze style snare line and hooks were set into position waiting for the fly-over hook or hooks to snag the parachute canopy for a successful aerial recovery. It was imperative that, in order for the aircraft commander to stay aligned with the parachute and payload, he must maintain an approximately 1,600-foot per minute descent at all times, through turns and straight flying, until contact and successful recovery. Then he had to immediately gain altitude to assure that the recovery line and the balance of the recovery gear would not come in contact with the rear horizontal stabilizer. In other words, once the recovery was in tow, the aircraft must not be below the towed object, it had to be above the object to assure safety clearance.

Our aircraft with the recovery gear out and the yellow booms down in position would remind most people of a wasp in flight. Our recovery gear would also remind people of going "fishing." We had our line and hooks. Our poles were our fishing rods and our winch drum assembly was the reel. Sounds like a fun assignment.

Sometimes our practice missions would be out over the West Coast at the Pacific Ocean or over the Salton Sea in southern California. This gave us early experience with over-the-water recovery training, which was to be our later destination, on and around the Hawaiian Island chain.

Later, as our aerial recoveries were becoming more efficient, we were briefed about the main purpose of all this special training and experimentation that we were going through. The aerial recovery of capsules returning from space (no pun intended) seemed "out of this world!"

Since our ultimate goal was to make aerial recoveries of these so called "outer space" capsules, someone suggested that we should be making experimental recoveries with something that at least looked like,

C-119J of the 6593d with its recovery gear extended during a training parachute recovery in 1959./Photo credit: USAF

and had a comparable weight, to a real space capsule. So the formed concrete blocks that we were using soon became obsolete as our new orange "dummy" capsules were introduced to the skies in our recovery areas. We were also introduced to some new problems when we caught these new payloads and had them in tow behind the aircraft. We soon realized that the package had excessive oscillation as it neared the rear of the aircraft. In the turbulence, the payload would whip around and oscillate, not just from side-to-side, but also up-and-down as well, and almost at times hitting the rear horizontal stabilizer of our aircraft. As our winch operator, Louis F. Bannick inched the capsule closer to the rear of the aircraft and into an area of less turbulence, it would settle down enough that we were able to reach out and manually help pull the object inside the rear of the aircraft.

This new capsule payload had much more drag tension than we had experienced with any of our other previous practice objects. It looked beautiful on the descending parachute, and they were both fairly stable together as they dropped, but when they were in tow, they were a dangerous situation. It was obvious that due to the extra manual efforts to pull the capsule into the aircraft, and considering all the new dangers, that it posed a threat to the aircrews and the aircraft. Some immediate design changes had to be implemented to reduce the excessive drag to an acceptable level. Further research and development, in conjunction with the Lockheed Missile and Space Division, the All American Engineering Company, and the aerial testing within our 6593d Test Squadron (Special), resulted in an acceptably stabilized modification of the capsules that assured, to a great extent, a smoother and safer tow-onboard recovery technique.

Our overall training programs were upbeat and demanding through our three months of temporary assignment at Edwards AFB. If we were not in the air performing our drops and recoveries, or on the desert floor doing retrievals of payloads and chutes, we were in special classroom training programs led by the All American Engineering Company that involved complete courses in the theory and techniques of aerial recovery operations. With these courses behind us, we were then able to disassemble, troubleshoot, repair, and reassemble the complete mechanisms of the aerial recovery gear.

Every aircrew had to meet a high qualification rating on recovery techniques and actual midair recovery successes. Other qualifications required by aircrews were the successful recoveries of simulated space capsules that had to be plucked out of a raft in the water, and also other capsules that had to be plucked from the ground, then towed onboard the aircraft while in flight. This training was necessary, at the time, in the event that a space capsule would be missed by an actual midair recovery attempt. We would then have the experience to quickly retrieve the capsule from the ocean surface or a remote land area, thus assuring that the capsule would then be flown to a final destination to have the data interpreted and analyzed as promptly as possible. Prior to either ocean surface or remote land aerial pickup by our aircrews of a *true, actual* capsule from space, it would be necessary that pararescue jumpers be dropped from aircraft other than ours to place the capsule in position with other types of gear so that we could snag and fly away with the trophy capsule.

Each aircrew was assigned as a regular stay-together team. Although, individually, we did occasionally fill in for a crewmember on another team who could not fly due to illness, emergency leave or some other legitimate reason. This gave us an opportunity to assist and also observe what some of the other aircrew recovery techniques were like. This allowed us to both see and share information. It ultimately provided aircrews that were not only proficient as regular teams, but also proficient as a unit, and could function well together.

Hill was assigned to C-119J #037 as a loadmaster. He explains his responsibilities as well as those of his fellow loadmasters. Hill gives an in-depth view into the work of aerial recovery.

Each recovery aircrew consisted of four loadmasters who were trained to do the same functions and were capable of performing all the duties required to properly set up the rigging of equipment, and upon a successful catch, bring the payload into the aircraft after it was winched to the rear of the aircraft.

Two loadmasters were assigned to each side of the rear of the aircraft. The poles were too heavy, and or clumsy for one person to safely and properly handle, so both loadmasters would unstrap the pole from its secured position on the inside of the fuselage and insert the smaller outer end of the tapered pole into the hydraulic actuator. The larger front of the pole could be held or the front loadmaster could secure the pole by strapping it to the side of the fuselage again, thus allowing both loadmasters to quickly rig the recovery line assembly and hook it to the rear of the pole. Then, at the time to extend the recovery gear out the back of the aircraft, the two loadmasters at the forward end would simultaneously ease and advance the poles out through the hydraulic actuators, while the other two loadmasters at the very rear of the aircraft would simultaneously, but gently, feed the center recovery harness and hooks out the center rear of the aircraft so that all of the recovery gear was in trail position, four simultaneous duties. Then one of the rear loadmasters would secure the line that was attached to the winch/trough assembly and the outer recovery harness into the emergency line severing system for safety reasons.

The last remaining duty prior to a recovery attempt was for a loadmaster to lower the poles to a downward position by using the hydraulic actuators, upon the request of the aircraft commander. During

A C-119J winch operator reeling in the Discoverer 17 capsule on November 15, 1960, Capt Gene Jones is at the right./Photo credit: USAF

the recovery attempt, a loadmaster used a headset to communicate with the aircraft commander about the estimated position and distance of the parachute if the recovery gear missed it. If the attempt had a successful contact, this would be communicated as well.

Once the payload was successfully snagged and in trail, the rear end communications to the aircraft commander was through the winch operator. As soon as the payload was in tow, and all was clear, the hydraulic actuators were activated to raise the poles in trail position, and then two loadmasters on each side of the aircraft would retrieve the poles and secure them to their stowage position on the fuselage. With the poles now out of the way, the payload was winched to the rear of the aircraft. Two loadmasters would take positions at the very rear of the aircraft and the other two loadmasters would position themselves near or close behind to assist in pulling the payload on board.

Most of the time, the same two loadmasters would work on the same side of the recovery operation. However, we would rotate the front to rear positions from flight to flight. When we started the air recoveries at Edwards, I worked the right side of the aircraft, which to us loadmasters would be to the left side facing our recovery gear out the back of the aircraft, and I remained on that side throughout most my recovery career.

Injuries to the loadmasters and aircrew were very minimal during our practice of midair recoveries and other training at Edwards. Although the duties performed by the aircrew put us in a highly dangerous environment, we quickly became very skilled in our performances, which resulted in, as I recall, only minor scrapes, cuts, bruises, burns or blisters (band aid-type fixes). Occasionally, we would receive a rope burn to our hand or fingers when letting the center, two-hook recovery harnesses out the back of the aircraft. This only happened if we let the half-inch nylon rope slide through our hands and fingers too fast and then attempt to slow it down by applying pressure, resulting in the friction burns. This would not happen, most often, if we wore our issued flight gloves.

The blisters would happen sometimes from excessive rope splicing. We were required to splice some "loops" when making our nylon rope recovery line harness assemblies. Every line-to-line connection and each recovery hook-to-line connection were all attached together with splicing methods. The hooks were put into the loops, and then the rope ends were spliced into the main rope, assuring a secure permanent attachment of the hooks into the loops. Placed inside each loop, prior to splicing, was a smooth, half rounded metal reinforcement that fit firmly and served as a cradle for the rope loop. The line end from the winch and the G-force reduction trough were attached to the lines of the outside recovery harness by a metal clevis that was inserted through the spliced reinforced loops. This attachment was then in position outside the rear of the aircraft and just beyond our safety firing mechanism. (At times when we had our "splicing-get-togethers," we would think that we were in the Navy rather than being Air Force recovery aircrew members.) The bottom line is, looking back forty-five years, thinking about our injury rate and all the dangerous situations that went with the entire aerial recovery program, we were very lucky!

Even with seasoned veterans training the new 6593d, there were several challenges to overcome. Hill explains four major issues he and other crew members had to work through.

While training to be experts in the space capsule aerial recovery business, there were many discouraging situations that came into play that were not reflected on the aircrews' performance or proficiency. Quite often we would return back to Edwards AFB, still disgusted with the events that took place during our practice midair recovery attempts for the day. Our squadron records that were maintained during the three months of training in late 1958 clearly indicate a summary of equipment failures that took place and were viewed by the aircrews performing midair recovery attempts.

The parachutes were the main recovery problem, accounting for fifty-five failures. Based on past experience and the initial drops, it was readily apparent that the first parachute systems furnished by the contractor, Lockheed Missile and Space Division, were not up to par. The stability and rate of fall were not satisfactory, reinforcement was inadequate, and subsequent testing confirmed this opinion. Stability and rate of fall tests were conducted to assess these characteristics with the Irwin Air Chute technicians. Out of these tests, several modifications were incorporated into the basic design.

Major alterations were done to improve the recovery parachutes. Two 1,500-pound test lateral reinforcing nylon strips were added to the parachutes, one-third and two-thirds of the distance from the canopy skirt. The entire canopies were reconstructed from single-weight nylon instead of two different nylon weights. Alternate suspension lines were increased in strength from 375-pound break strength to 1,500-pound break strength. Parachute canopies were increased almost 3 feet in one model, in another model the canopies were increased almost 4 feet. The suspension lines were also lengthened 6 inches in both models of parachutes. Other minor changes were also tested and incorporated. There were also two different parachute models. One model (Mark I-3) had a near 17-foot canopy that could support a maximum payload weight of 60 pounds. The other model (Mark II-3) had a near 24-foot canopy that could support a maximum payload of 120 pounds. Acceptable stability and rate of descent, 1,600-1,700 feet per minute, along with satisfactory reinforcement were obtained, and the decision was made to use this system on operational recovery missions.

The second leading cause of recovery failures were line and/or hook mishaps (for a breakdown of recovery failures or "misses" by type and amount, please see the Recovery Statistics table below). Most of these were line failures due to the G-forces. The corrective measures taken were discussed earlier relating to the type of line changes and the addition of the G-force reduction trough equipment. A few failures were due to hooks and this was also basically eliminated by the G-force-reduction factor.

The excessive rate of oscillation was the third highest cause of recovery failures. This was also discussed earlier, where changes to the dummy capsules were made. The parachute improvements mentioned were a factor as well.

Winches and poles were the fourth highest contributing factor of recovery failures during our practice midair recovery attempts. Occasionally, a winch would cause the failure of an aerial recovery effort, due to a braking system pressure fault. We would then lose the payload because of the sudden G-force caused by too much brake, or lose a payload because of no brake at all (this was rare, but corrected early in the program). I recall seeing, overall, several bent poles that were bent at time of a recovery contact, but I don't recall broken poles.

I remember one time returning to Edwards AFB, from a practice mission, with a bent pole remaining partially outside the rear of the aircraft. The landing was done with the rear ramp open, because the bent pole could not be pulled back through the hydraulic actuator. I know of no standard replacement plan of the poles due to metal fatigue. Naturally, the bent poles were replaced. I feel that one particular reason that we did not have more damage to the poles was due to their design. They were manufactured with a taper from end-to-end, or let's say front-to-rear, allowing the smaller diameter end at the rear to be placed into the hydraulic actuator and then pushed to the outside of the rear of the aircraft. When fully extended outside of the aircraft, the front end (or largest diameter end) of the pole would be firmly seated and secured within the actuator. Simply put, the outermost ends had more flex, as does a fishing rod.

The table listed below contains a brief summary of the statistics of our practice air recovery missions at Edwards AFB, from September through the first week of December 1958.

Recovery Statistics	
Total airdrops	272
Total recoveries	157
Total missed	115
Percent of recoveries	57
Justification of misses:	
Parachute failures	55
Line or hook malfunctions	23
Excessive oscillations	12
Winch or pole malfunctions	6
Total failures or malfunctions	96

As indicated earlier, when the failures that were related to equipment were deducted from the statistics, it was expected that the 6593d Test Squadron could maintain successful midair recoveries of approximately 87 percent.

Hill describes the differences between midair and ground or water recoveries. For the loadmasters and winch operators, the process was the same. It was the pilot who had to adjust his flying for the different types of recoveries.

Ground and water recoveries were as much a challenge to the aircrews as the descending parachute midair recoveries. Perhaps they were even more of a challenge. The equipment was set up by a qualified surface rigging team. A lightweight pole (a pole similar to one used to support a volleyball net) was attached to a supporting base, allowing it to stand freely. This assembly would sit on any relatively flat surface allowing the pole to be in a vertical position. A half-inch nylon rope line was attached to the recovery capsule that was placed in position, either on the ground or on a raft. The line was then placed up the 8 to 10-foot pole and attached to a festooned harness at the top. Attached to the line harness was a fluorescent orange-colored flag that was set at the very top of the pole. With the surface rigging being a simple process and complete, it was now up to the aircrews to approach for a surface-to-air recovery attempt. Both the bright yellow raft and the bright fluorescent orange-colored flag provided a good visual target.

Our recovery equipment and gear for surface-to-air recoveries was the same as we used for midair recoveries. From a winch operator's and loadmaster's point of view, our recovery setup and recoveries

C-119J about to recover a floating training capsule in 1960./Photo credit: USAF

Two C-119J loadmasters hauling in a data capsule recovered from a ship in 1960./Photo credit: USAF

were identical for either type of recovery. The aircraft commander had a different perspective, in that, in lieu of chasing a descending parachute, he had to assure a line up on the brightly-colored objects on the surface. He had to maintain a very smooth, level approach. His recovery contact pattern being extremely small, should he be slightly high, it was a near miss. Should the aircraft be low to some extent, we could lose our recovery gear from our poles when the first hook hit the ground or the water surface. Either way, we lost a recovery if he was not on a perfect approach.

During a good approach, and after contact was first made with the stationary harness, the G-forces would provide a clean break away from the vertical pole. The harness and the bright fluorescent orange capsule in tow would quickly be taken to a higher altitude to be towed to the rear of the aircraft and pulled on board. The quick ascending altitude was a must to prevent the capsule from hitting the ground or the water, thus avoiding damage to the capsule or suffering a lost recovery.

The aircraft commander would be challenged again when performing water surface recovery attempts. Depth perception would be different due to the lack of ground objects, and lining up on the target could vary based on how choppy the water might be at the given time of the recovery attempt. Our aircraft commanders weren't jet fighter pilots, but they were perhaps better when flying the recovery missions in their own skies.

Being an aircrew on the receiving end of the recovery packages, did not allow us to get totally familiar with the ground/water surface rigging. We saw the rigging at the time of recovery and what we pulled into our aircraft. We did have the opportunity several times to be on the surface and observe some of our other aircrews perform surface-to-air recoveries. It was very interesting and exciting. I wish that I could remember more precisely the exact rigging, so that I could explain it in detail, but again, that was several observations that took place forty-five years ago.

I had mentioned earlier, Harry Conway, the All American Engineering tech rep, was a trusting and daring individual. You may have seen the film showing a man in a harness being picked up from the ground by an aircrew with recovery gear, if so, that was probably Harry.

Most often, the ground-to-air recoveries that I remember, the practice recovery capsule was placed in a bright yellow raft that sat on the ground and all the rigging was inside the raft, same as a raft/water surface recovery. The recoveries were performed this way to provide a vividly-colored object to aim at, as the aircraft commander homed-in on the bright fluorescent orange-colored contact flag.

In late 1958, we flew a practice recovery mission and needed to fly down to El Centro Naval Air Facility, California, to pick something up. Being the lead aircraft for A Flight, Capt Mitchell guided our A Flight over the El Centro facilities flying in a tight formation, did a series of tight formation turns and maneuvers, then peeled off in a sharp descending turn with the other aircraft following directly behind for landings. The pilots taxiing in a precision manner brought the planes to a stop near the base flight operations, fully reversed the props, and backed the aircraft into uniform parking positions on the flight ramp. Walking to the operations building we passed several of the Blue Angels pilots who happened to be there. They were amazed at our pilots' "outstanding" C-119J Boxcar performance, indicated so, and advised several aircrew members as we passed that the "head" (latrine to us) was just inside if we needed to clean ourselves up after riding in that performance. We all had a good chuckle. They knew and we knew that we had some of the best pilots in the Air Force.

After training at Edwards AFB, the 6593d moved to their permanent station at Hickam AFB, Hawaii. Hill summarizes his training then explains the process to get everyone (including the airplanes) to Hickam. He also notes that the citizens of Hawaii quickly came to know the new squadron and the sound of their C-119s.

By early December 1958, when our temporary training assignments were complete at Edwards AFB, our test squadron had successfully completed 157 practice midair recoveries. We had 115 total misses, resulting in a 57 percent overall proficiency. However, these figures do not reflect an accurate degree of aircrew proficiency, because many air recovery attempts were performed with sub-standard parachutes that were seven years old, and numerous air recovery attempts were not successful during the contractors' systems modifications testing. After ruling out failures caused by parachutes, excessive oscillation of payloads, winches, lines, booms, and hooks, it was concluded that the flight crews had achieved a high degree of recovery technique reliability in the air. At that time, it was expected that the 6593d Test Squadron could maintain successful midair recoveries of approximately 87 percent.

It was previously decided that one of the C-119s and its aircrew would remain at Edwards AFB to accomplish additional research and development, and complete the training of the water recovery techniques. This lone aircraft and aircrew were estimated to then arrive at Hickam AFB by February 1, 1959.

Chapter 2 - A2C Daniel R. Hill

During the period December 2-10, 1958, all but one of the C-119s departed Edwards AFB for Hickam AFB with a stopover at Travis AFB prior to the overseas adventure. Onboard these flights were the aircraft commander, copilot, navigator, flight engineer, and the aircraft mechanic. The winch operator and loadmasters did not go on the long flight over the ocean from the mainland to Hawaii in the C-119s. We flew on a Military Air Transport Service (MATS) C-121 Constellation passenger aircraft with the balance of the squadron personnel and their dependents. We had no function to perform while ferrying the C-119s to Hickam. The less weight the better. As a rule, our C-119s with full aircrew, recovery gear and extra fuel were approximately 4,000 pounds overweight.

Just prior to the long overseas flight, four of the C-119s still needed to have the extra large fuel tank installed. A Fairchild Aircraft engineer arrived at Edwards with blueprints to supervise and assist the installation of the Benson long-range fuel tanks and tank enclosures. The proposed time for installing the equipment on each aircraft was eighty hours and forced a scramble to complete and test the tank installation. These large tanks would hold 1,000 gallons (I believe) of additional aircraft fuel that would definitely be required for the long flight to Hawaii, and also on subsequent flights on capsule recovery missions out over the Pacific Ocean at later dates. The enclosures were made of wood framing covered over with thick plywood to basically hide the tanks. They also served as a flat surface to stow supplies such as parachutes, flight lunches, prefabricated recovery harnesses and miscellaneous items. The whole tank assembly sat several feet behind the bulkhead near the cockpit and extended to just forward of the recovery winch drum assembly and the winch operator's position. Also, prior to the overseas ferrying of the Flying Boxcars, the maintenance personnel performed their periodic inspections on all the aircraft, even though on some of them the required time was not yet due. During this time, they found it necessary to replace a total of four engines.

The first 6593d Test Squadron (Special) personnel arrived at Hickam AFB on December 3rd. They were some administrative airmen, followed by the main body of the squadron that arrived December 4-10, 1958. By December 15, the recovery aircrews of the 6593d Test Squadron (Special) were now relocated and operational at Hickam AFB. We had been assigned a key role in the new Discoverer space program.

Through the efforts of everyone in the 6593d Test Squadron (Special), the period December 2-10, 1958, was devoted exclusively to preparing for and accomplishing the main movement of aircraft, flight crews, equipment, and the balance of the personnel from Edwards AFB to Hickam AFB. All arrived without incident. However, shortly after arrival an engine had to be replaced on one of the aircraft due to the wear and tear of the long overseas flight. Overall, it was not a bad performance for a bunch of guys from Pope, Shaw, Sewart, and Ardmore AFBs that had formed a new squadron, became fairly proficient in several months of research and development, and moved on to the Pacific Air Command in Hawaii.

In a short time we were introduced to our new quarters, both living and hangar facilities. Several days were allotted to familiarize ourselves with our surroundings on base and to organize our equipment within our newly assigned squadron environment. By December 15, the aircraft and recovery aircrews were in place and operational at Hickam AFB. From December 15-31, all our aircrews accomplished at least two successful practice midair recoveries and were judged to be 100 percent qualified.

Most of our heavy, lumbering C-119 take offs were toward the mountain passes. We flew a rather steep right turn that would line us up parallel to the coast off of Honolulu, Waikiki Beach, then past Diamond Head. We then set a course for our ballpark practice aerial recovery area.

C-119J loadmasters during a 1959 parachute recovery./Photo credit: USAF

> "It was not long until the other military personnel stationed and the civilian population in Hawaii knew the sight and sound of our C-119 aircraft."

It was not long until the other military personnel stationed and the civilian population in Hawaii knew the sight and sound of our C-119 aircraft. Because of an earlier announcement (government statement and press release, December 3, 1958) that proclaimed the start of the Discoverer program, and advised about the launching of satellites at the rate of one per month starting in January 1959, the inhabitants on the ground below us knew that the "nose cone catchers" were here and active in the skies around them. The news media dubbed us that handle pertaining to the space capsule recoveries.

Into January 1959, hours, days and weeks became saturated with preparing for and practicing recovery missions. Drop and catch, drop and catch were our steady aerial recovery routines. When we were not in the air, much of our duty time was utilized inspecting our rope lines, rope harnesses, hooks and personal equipment. Depending on the condition, if it wasn't next to perfect, it was mostly replaced. Refurbishing was almost out of the question. We spliced and assembled many brand new line and harness recovery kits. Our free time was minimal at first. The continuation of packing the parachutes that we used for practice recoveries was an ongoing duty, as was keeping our assigned aircraft in a clean, neat and orderly condition. We did not get to see much of Oahu Island the first several weeks, except for what we would see from the air.

It was not just the recovery aircrews, the entire squadron was hustle and bustle. The administrative, supply, maintenance, and the whole outfit were highly motivated. We loved it and we knew that we were the talk of the town, on and off base. On base, when we would be going from place to place, it was often overheard, "There are those capsule (or nose cone) catchers." Anxieties were growing strong; the Discoverer program was very near its first historic liftoff.

Preparing for a mission was detailed and dedicated work. Not only were the 6593d crews on alert once the satellite was launched, they had to verify and inspect all of their equipment. Hill explains the steps he took to get his airplane ready. After all their recovery training, every crew member was excited about flying an actual mission.

Our recovery preparations for the "live" Discoverer recovery attempts were somewhat different. We were always serious when preparing and checking all of our recovery aircraft and equipment, whether it was a practice or a live recovery attempt, but there was finer detail and closer observance placed on the equipment being prepped for a real Discoverer payload recovery. As stated before, we performed numerous practice recoveries, so anytime that we were scheduled for a live capsule recovery, we always replaced our used half-inch diameter nylon winch rope with a new 700-foot one. We assured that it was wrapped on the winch drum very tightly to hopefully avoid any possible line overlays or crossovers. Also, we would replace our used half-inch diameter nylon rope harnesses with hooks attached, to assure that we had the best quality material in place for the live recoveries. By doing this, there was one possible downfall to this mode of operation; the new line, line harness, and hook assemblies had never been tested during a practice capsule recovery. Any way you look at it, the odds were still favorable by not taking a chance with used gear that could possibly be nicked or frayed.

As a rule, in addition, the winch operator and the four loadmasters of each aircrew would handle and visually inspect every inch of our new line and assemblies to further assure that top quality materials would be utilized. The aircraft, recovery equipment, and our personal gear and clothing were maintained in a clean and neat manner, thus assuring and reflecting a good image of the aircraft and aircrews should we return to Hickam AFB with the first historic aerial recovery of a Discoverer space capsule, and meet the high ranking officers, news media and spectators.

Sometime in 1959, some of us loadmasters volunteered for an exercise program that started with extensive swimming training. The thought behind this plan was to begin an initial phase in becoming pararescue qualified. Once we were in good shape and passed the requirements of the swimming program, we would then be scheduled to go to jump school training and eventually be qualified pararescue personnel. Then, should any space capsules splash into the ocean (that had somehow avoided our recovery gear), we could drop a life raft out of our C-119 near the floating capsule, parachute into the water, swim to and rescue the capsule, and then put it into the raft. It was also considered that we could rig the capsule for a water-to-air pickup by one of our squadron planes for a quick return of the capsule back to Hickam. Unfortunately, as we were nearing the completion of the swimming training program, the para-jump training exercises were canceled. We were told that the training was halted because there were pararescue personnel already trained and available. This meant that we would stay in the air.

I'm not absolutely sure of the total timeframe given, but for every Discoverer launch we were advised about it, and we would be placed on standby alert. What I mean by that is, once we were informed about the planned liftoff date, and if we were not flying a practice recovery mission, it was the responsibility of all the aircrew members to consistently check in with our 6593d Test Squadron Operations. At our assigned hangar and flight operations there was a sign-out board with all our names on it. Our instructions were clear and very simple; it was mandatory that we wrote the precise location, address and phone number

The interior of the Discoverer 12 capsule before it was launched on June 29, 1960. Discoverer 12 failed to reach orbit after its launch./Photo credit: USAF

where we could be contacted. When we left the location that was previously noted on the board, we had to immediately report back to our flight operations and post our revised location and other information on the sign-out board. At all times the flight operations administrators knew exactly where we were, and all this was necessary, so that if the planned launch was moved up, so were we. Even after a launch, the planned recovery orbit may have needed to be moved up on the schedule, meaning that the aircrews would have to fly out to the ballpark and outfield recovery areas earlier than planned.

All of our aircraft and aircrews flew on all the Discoverer recovery attempts. Our awareness of a Discoverer launch was not a drop of a hat situation. I would say that we were placed on alert several days prior, even while we still performed practice recovery missions.

Being so long ago, I am not positive of how many Discoverer recovery deployments we participated in, based on the malfunctions of either the first (Thor) or the second (Agena A) stages of the launches or orbit entries. Reviewing the history of Discoverer 1 through 12, only half (six of twelve) of these launches made it into orbit, and only three of the twelve times did the capsule separate, of which only two reentered earth. Should a failure occur with the launch or in space prior to a particular designated time, our Discoverer recovery deployments were called off.

The number of orbits the Agena A completed determined whether we would fly out to our recovery areas or abort the midair recovery attempts. Normally, if the Agena A reached its fourteenth north-to-south polar orbit, we would be given the green light to fly out to our recovery areas. With the spacecraft in its fourteenth orbit, this would allow us time for final briefings, double-checking our equipment and gear, and then to go for several-hour flights from Hickam AFB and be in our designated recovery areas one hour prior to the scheduled capsule reentry back into the earth's atmosphere.

Looking back in time and also reviewing the historic data of Discoverer 1 through 12, I am estimating that our recovery aircrews of the 6593d Test Squadron (Special) would have deployed to our capsule recovery areas at least five different times for possible live recovery attempts. These would include Discoverers 2, 5, 6, 8 and 11. Discoverer 2 was the first to achieve orbit, the capsule ejected on the seventeenth pass as planned, but unfortunately it was nowhere near our recovery areas and it was said to have come down near Spitzbergen. The retro-rockets on Discoverer 5 fired with bad orientation, raised the orbit, and the capsule was not recovered. The retro-rockets malfunctioned on Discoverer 6 and there was no vehicle separation. The Agena malfunctioned on Discoverer 8; it had a high orbit, separated on the fifteenth orbit, overshot our 36,000-square mile recovery areas, and the chute failed. Needless to say there was no recovery. The recovery vehicle on Discoverer 11 ejected, the spin rockets exploded, and the recovery vehicle was not recovered.

Let's compare these "experimental" Discoverer launches to something that perhaps most professionals can relate to in normal life situations. Imagine having studied, then greatly practiced a given objective until you were considered an expert or a professional in that given skill, then the time finally came to put your efforts into play. You had long hours and months of preparation; anxieties have been at an all-time high, nerves have been racked and a pressure continues to build within you. "Stressful" may not be the right terminology, but it was overwhelming. You have traveled many hours to get there and you are proudly prepared to do your thing, then you discover that no one shows up on time, and they won't even let you know if they are going to be late or if they are not coming at all. You find yourself stretching and straining your neck looking for a contact to appear. Finally, after what seems like a long anxious period of time, you realize the big letdown. You know that it isn't going to happen at that point. So you make the best of the situation and continue to practice and practice for the next unknown opportunity.

The anxieties remained high in our 6593d Test Squadron (Special), even though many of the Discoverer payloads did not make it into orbit. When this happened and our recovery deployments were scrubbed, we were still energized to get out there and have the chance for what we had been training to do.

Flying out to our practice recovery areas was not stressful. It was a routine job that we very well loved. We really looked forward to performing the missions, and enjoyed the practice aerial recoveries. To most of us, just to have this opportunity was a great excitement. Being out there was gratifying. Most of the time, it was a letdown of our spirits when we knew that a mission was coming to an end for that particular day as we neared Hickam AFB on our return trip home. For months upon months we practiced the aerial recoveries and never once got tired of doing them.

> *"To most of us, just to have this opportunity was a great excitement. Being out there was gratifying."*

By the launch time of Discoverer 2 on April 13, 1959 (the first to go into orbit), our squadron had seven months of practicing aerial recoveries behind us. By the launch of Discoverer 11 on April 15, 1960, we had over 19 months (or over 1.5 years) worth of practice recoveries under our belts. By the liftoff of

Discoverer 12 on June 29, 1960, we were edging up on 22 months (or nearing two years) of credit devoted to what we did best, practice.

One big advantage that we always had with the performance of our practice aerial recoveries was always coming home to Hickam with the midair dummy capsule recoveries on board our aircraft. To achieve that with a live Discoverer capsule remained to be seen.

To see a live Discoverer payload from space descending towards you on a brilliant parachute was every aircrew's dream from day one. Discoverer recovery deployments were the biggest thrill and objective that all the aircrews of the 6593d ever had. Perhaps the biggest letdown and defeat that we ever had was returning to home base with not even hearing a beep or having a glimpse of a Discoverer space capsule on a parachute.

At this time in history, there were several other space activities during the Cold War years. The Discoverer program, including our aerial recovery phase of it, was a well-known fact, not just to those who worked with it, but to the news media, our folks (family) back home on the mainland, and to most of the world itself. The media would broadcast or print the news of every Discoverer launch. Worldwide attention and the folks back home waited to hear or see just how our recovery aircrews performed in an effort to catch a falling space capsule. So, it is almost needless to say how disappointed we were to let most of the world down at such an important time in our aerospace history.

Something had to change, and change very soon. We needed a Discoverer space capsule to eject on target and enter the earth's atmosphere over our recovery area.

Hill is not shy to admit there was intense competition among the 6593d crews. He describes the friendly banter, bets, and good old fashioned teasing. The competition boosted morale and pride among the aircrews in the squadron.

"Competitive" may be a mild word for most of our 6593d Test Squadron (Special) personnel. Let's try "dog-eat-dog" competition, although it was "kept in hand," so to speak, since we were good friends and worked well with one another. Our aircraft commander, Capt Harold E. Mitchell was once quoted in an interview as saying that the 6593d had no sympathy dispensers.

There were not too many dull moments. If things seemed to get a little quiet, someone would instigate by needling or gibing the others to get something started. This was not only in the enlisted ranks, but applied by officer-to-officer as well. Ground crew-to-ground crew and flight crew-to-flight crew, the agitations lingered. TSgt Louis F. Bannick, the winch operator on our aircrew, aircraft number "037" (#18037), had the reputation as being probably the biggest agitator of getting something started to hassle his counterparts. It was a joy to always observe what Louie would come up with next. He also kept most of the officers of the 6593d on the lookout.

> "He said if I carried this silver dollar that it would bring me good luck and I would never be broke. I still carry that same coin in my wallet today, and I have had some good luck and I haven't been broke."

We even competed on who would buy the coffees that we drank when we were in our flight operations area (hangar). We had small groups of odd man out, until we got down to two people who would then flip

Chapter 2 - A2C Daniel R. Hill

A2C Daniel Hill (seated in the center) and some of the 6593d loadmasters playing cards in 1960./Photo credit: USAF

their coins (odd or even, or heads or tails) to decide which unlucky airman would pay for the coffees for all in the small group. I began using an 1881 silver dollar to flip for my gambling part of the competition. Charles "Chuck" J. Dorigan (a friend and loadmaster) previously gave me that particular coin. He said if I carried this silver dollar that it would bring me good luck and I would never be broke. I still carry that same coin in my wallet today, and I have had some good luck and I haven't been broke.

When things would get a little rough between someone from crew-to-crew, the balance of the crew would always come through for the rescue (moral support). At this time, the so called "moral support" was mostly agitation stirred up to harass the other crew as to who was the best, with a friendly threat as to which aircrew would be successful with the first air recovery of a Discoverer space capsule. There was not an aircrew that was not chomping at the bit to be the first, to be lucky enough to recover in midair the first live space capsule from a Discoverer launch. Bets and bragging were a prerequisite.

Hill's C-119, #037, was in the Pelican 1 position for the Discoverer 13 recovery. This was the best location to recover the capsule. He explains every aspect of that day. The entire crew had high spirits and was full of enthusiasm. Unfortunately, they did not catch the Discoverer 13 capsule.

Our aircrews of the 6593d Test Squadron (Special) had been placed on alert, and then Discoverer 13 was launched on August 10, 1960. The Thor booster placed the Agena A in position to achieve space orbit. As the orbits progressed, we anxiously waited for the mission deployment to our recovery areas over the

Pacific Ocean. The word was that the space vehicle orbits were going pretty well, and it was expected that we would depart at the selected orbit number to our designated recovery areas. If all went well, we would reach our assigned recovery area in several hours, which then would allow an extra hour to double-check and get our devices and recovery gear ready prior to the capsule ejection from space.

During the briefing it was announced that A Flight would be assigned to the ballpark area and that our aircraft (#037) and crew was given the code name Pelican 1, meaning that we were smack dab in the hot spot or the primary recovery location. In other words, our aircrew had the location of choice with the best chance of making the recovery, if all of the space flight and reentry systems were on track.

Hours passed as the Agena A with the recovery vehicle entered the designated space orbit on August 11, 1960. The track was good and we received our recovery mission deployment, so we aircrews boarded our aircraft and took off. As the island of Oahu started to fade out of sight, our aircraft commander, Capt Mitchell set a course to take us approximately 250 nautical miles west—northwest of the big island of Hawaii.

The spirits and hopes of everyone in our aircrew were high. You could just feel it. As we lumbered through the skies in our overweight C-119J, we felt a sense of joy and assurance that today may very well be the day that we had worked so long and waited for, to get a chance to have a nibble at the very first live Discoverer space capsule to enter our recovery area. After all, we were headed for the driver's seat of the recovery operation, the prime projected recovery slot.

After our arrival, we checked and double-checked our equipment and recovery gear. We were now engaged in a waiting game for the capsule ejection, separation, reentry, and signal. We were all ready, with anxieties going rampant, just the thought of our big chance and the position that we were in on this particular recovery mission. One could almost taste the victory and glory, to be the historic first. Not just for ourselves, but for the rest of the 6593d Test Squadron, all the people involved in the Discoverer space program, and the balance of the world population that was rooting for us to be successful on this afternoon's recovery mission. I never really asked any of the rest of our aircrew, but I feel almost certain that the ten airmen inside that noisy Fairchild Boxcar aircraft uttered a prayer or two that all systems would go well, and that we would all get our sights set on a descending space vehicle on a parachute.

On the seventeenth orbit it was time for the Agena A to be placed in position for the capsule separation. The retro-rockets would be fired to thrust the recovery package back into the earth's atmosphere. Below the Agena, our recovery aircraft, along with support from communications aircraft and water recovery ships were keyed and waiting. Providing that systems were still operating satisfactorily upon reentry, the brilliantly-colored parachute should be deployed around or under 65,000 feet. The chute and space capsule should be overhead descending toward our recovery areas.

Finally, the silence was broken on the recovery end of the Discoverer 13 operation. The Hawaii ground station advised that they had just picked up a signal from the space capsule, and also confirmed that the parachute deployment had taken place as planned. Somewhere thousands of feet below the descending package from space, our recovery aircrews, and support control aircraft, along with recovery ships started to pick up the radio beacon signals. I need to mention, at this point of the communications receiving end, that when our Pelican 1 aircraft navigator, 1st Lt Bob Counts, started picking up the space capsule's signal, it was very difficult to determine the precise direction at first.

Meanwhile, one of the RC-121 control aircraft was picking up a signal and radioed our Pelican 1 aircraft to target on a vector of 285 degrees. Upon flying this advised course for what seemed to be too long of a

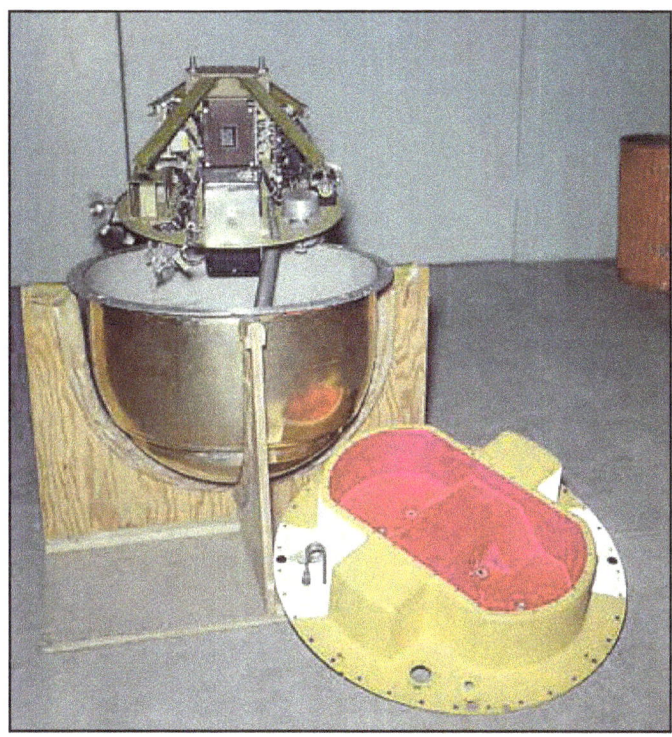

The Discoverer 13 space capsule in 1959 before it was launched on August 10, 1960.
/Photo credit: USAF

time, 1st Lt Counts requested our aircraft commander (Capt Harold Mitchell) to fly a 360-degree turn, to affirm the direction of a strong signal. While completing more than half of the turn, Bob concluded that the course given to us earlier by the control aircraft had directed us away from the correct recovery area. In fact, Bob was positive that the space capsule was descending behind us!

Immediately, Capt Mitchell reversed our course and flew as fast as possible back to the recovery area that we had originally just come from. Arriving at that beacon location from where we had started, we saw another one of our squadron aircraft arriving from another direction as well; it was Pelican 3, also of A Flight. To our recovery crews, it was extremely disappointing to see what we saw next. A silver and orange parachute was lying in the Pacific Ocean and attached to it, floating in the water, was the beautiful Discoverer space capsule. What was the most disappointing to our aircrew of Pelican 1 was the known fact that we were given a wrong vector by someone in the RC-121 control aircraft, which we now knew, took us away from where we should have been to perform our midair recovery. The floating capsule and parachute was directly below our planned Pelican 1 recovery area. It appeared that all the systems of the Discoverer 13 program worked and were on target, with the exception that our aircrew of Pelican 1 had been given erroneous information from a so-called "qualified" control aircraft.

To see the space capsule bobbing in the water brought many concerns about how long it would float prior to sinking to the depths of the Pacific. We knew from prior briefings that there was a slowly dissolvable plug in the space capsule to assure that the capsule would sink when it was saturated with salt water after an allotted period of time, to avoid another country (at this time, mainly the Soviets and their counterparts) from retrieving our USA-valued space property. We aircrews knew that our backup source

The Discoverer 13 parachute floating in the ocean./Photo credit: USAF

Loadmasters from the Pelican 1 aircraft dropping smoke markers near the floating Discoverer 13 space capsule./Photo credit: USAF

was Navy personnel qualified to drop into the ocean, swim to the capsule and attach it to a safety line and raft, if necessary, to assure security until it was either picked up by a helicopter or a ship.

At this point, not knowing exactly how much time it would take for the Navy to arrive, and not knowing for sure if the capsule would float the prescribed time prior to sinking, several of us asked for permission to drop one of our aircraft's inflatable survival rafts into the Pacific near the floating space capsule, and then for several of us loadmasters to parachute into the ocean and attempt to attach and secure the capsule for security and recovery. I must admit that it was a good try on our part, but needless to say, we were denied the permission, but not the possibility. As we flew near the floating space capsule we dropped dye markers and smoke bombs out the rear of our aircraft door to aid in spotting its location.

The aircrew members on both aircraft closely looked out of the cockpit windows, the side fuselage windows, or the open rear of the aircraft (whichever view was best at the time). Our Pelican 1 and 3 crews continued to fly patterns over the gold-colored space capsule and its silver and orange-colored parachute, as if to say, "We are watching and guarding you as you now shift on the high ocean waves with your yellow and green dye markers floating with you, as you send your strong strobe light signal to us, after your seventeen space orbits around earth."

After flying and sentry observing for what seemed to be an endless time (approximately three hours) since the Discoverer splashed down, a helicopter (deployed from the USN Ship, *Haiti Victory*) appeared on the scene to have its divers attach and retrieve America's first manmade object from space. Having observed a successful retrieval by the Navy, our aircraft and crews of Pelican 1 and 3, once titled "primary recovery aircraft" for the Discoverer 13 launch, had permission to depart the assigned recovery area. We flew disappointedly home to Hickam without our space capsule onboard our USAF C-119J.

> *"When we saw the capsule bobbing up and down with the estimated 12-foot waves, it was a glorious sight that our aircrew will never forget. But, on the other hand, it was a sight that could not help but to put anger, disappointment, and disbelief into the minds of all the aircrew of Pelican 1."*

When we saw the capsule bobbing up and down with the estimated 12-foot waves, it was a glorious sight that our aircrew will never forget. But, on the other hand, it was a sight that could not help but to put anger, disappointment, and disbelief into the minds of all the aircrew of Pelican 1. We all thought this first capsule was meant for our aircraft and aircrew since it came down exactly within our assigned ballpark recovery area. It had to be the biggest disappointment that our aircrew ever had concerning the entire Discoverer program.

To see this wonderful sight there below us as we flew over and over, as we performed sentry observations hour after hour, thoughts and dreams kept coming to mind as we looked back and thought about all those practice missions that were behind us. At that particular time it was the raw feeling of having salt rubbed into our wounds as the old saying goes. On the upbeat side, it suddenly dawns on you that what this object below you represents is not really a personal, aircrew, or squadron event, but a huge historical victory for the Discoverer program, the USAF, the Navy, all of the military, the entire USA, and the world as a whole. Granted, an aerial recovery achieved by the USAF would have had this capsule back to Hickam and on its way to Sunnyvale, California, for study and observation by the time the Navy helicopter crew retrieved it from the ocean, but not this day.

A Navy helicopter hovers over the floating Discoverer 13 capsule as Navy frogman Boatswain's Mate Third Class Robert Carroll jumps into the ocean to recover it on August 11, 1960./Photo credit: USAF

The Discoverer 13 capsule (within the metal container) being removed by Col Charles Mathison at Hickam AFB on August 12, 1960./Photo credit: USAF

Naturally, our entire squadron felt disappointment and defeat when the word was spread throughout about what had taken place on the day of August 11, 1960. But also, the whole squadron knew that the most important thing with the Discoverer program, at this time in history, was to successfully recover a capsule that had been in space. Our non-midair recovery of the Discoverer 13 space capsule was only a detour in history for our aircrew and the whole 6593d Test Squadron (Special).

It was later believed that we were directed toward a thunderhead when we should have had the option to stay in our prime recovery area. It was also determined that since the space capsule was descending on target directly above our Pelican 1 aircraft, the beacon signal that our navigator, 1st Lt Bob Counts, had difficulty on picking up right away was saturating our aircraft receiving equipment. This saturation supposedly caused an imbalanced and distorted tracking signal for that short interval of time.

I can remember, at the time of the recovery, reading about our recovery efforts on August 11, 1960 in the Honolulu newspaper and also seeing something on the local TV. They described how a helicopter crew from the USN ship, *Haiti Victory* retrieved the Discoverer 13 space capsule. A frogman jumped into the Pacific Ocean then swam to the capsule, attached it to a line from the helicopter, and the capsule was then safely hoisted onboard. Then the helicopter crew hoisted the seaman onboard and flew back to the ship.

At that point in history, I would think that the Navy would have celebrated to some degree. Perhaps they flew the helicopter crew, or at least the seaman who jumped in the ocean and attached the capsule to the hoist line, back to California and then on to Washington, D.C., for some type of a medal awards and

The Discoverer 13 capsule when it was delivered to Andrews AFB in a C-130A on August 13, 1960. Left to right: Lt Gen Bernard Schriever, Gen Thomas White, and Col Charles Mathison./Photo credit: USAF

recognition. It certainly deserved something like that. After all, as we all know, this was a very big break for the Discoverer program and the space race. It was also a victorious event as well as a huge historic first for America.

Even though we observed from the air, I would like to meet or talk to the seaman who attached the recovery line to the capsule. It would be interesting to hear his personal experience of that significant performance in the historical chapter of the Discoverer program. References list his grade and name as Navy Boatswain's Mate Third Class, Robert W. Carroll. He and the Navy must be very proud of their accomplishments.

After missing Discoverer 13, Hill and his aircrew were ready to go out again on another mission. They were rotated to the Pelican 9 position which had the lowest chance of recovering Discoverer 14. Regardless, they prepared their C-119 for the mission. Hill recounts every detail of that day.

As indicated before, we aircrews were always placed on alert prior to any Discoverer launch, Discoverer 14 was no exception. In preparation for a possible recovery attempt, each aircrew very carefully inspected the equipment and recovery gear to assure their best quality condition prior to getting deployment approval for the recovery mission. TSgt Lou Bannick, our winch operator, supervised as we loadmasters assisted with the removal of the winch line that now had seen several practice recoveries since the Discoverer 13 mission a week prior. Our standard operating procedure prior to any live Discoverer recovery attempt was to replace the used line with new line. Very slowly and carefully we inspected every inch of the new 700 feet of half-inch nylon rope as Lou operated the winch. Two of us would keep the line weighted while the other two would assure that the new line was wound on the hoist even and firm and ensured that no kinks or overlapping existed. Every recovery mission that we flew, the equipment and gear was inspected, but on live Discoverer recovery attempts, it was inspected, re-inspected, and then some, as you will see.

The Discoverer 14 liftoff from Vandenberg AFB was scheduled for August 18, 1960. The liftoff of the Thor was shortly after noon, near 1:00 p.m. A successful orbit was obtained by the Agena A with the capsule vehicle attached. As our 6593d Test Squadron (Special) received word of the good condition of the spacecraft, it put us in a position of readiness and waiting (time to check out our recovery gear again).

Finally, as the spacecraft entered its fifteenth successful orbit, we aircrews were given the go-ahead to depart from Hickam and fly a southwest course to our assigned recovery positions. At the earlier briefing of the aircrews, we learned that our aircrew #037 was assigned the code name of Pelican 9, which meant that we would be in the lowermost position of the recovery effort. This was the exact opposite and the least desirable recovery position, especially compared to the week before during the Discoverer 13 recovery mission attempt when we had the best and most desirable recovery position. To have the least wanted position for the Discoverer 14 recovery attempt really came as no major surprise to us, due to the fact that the positions rotated from recovery mission to recovery mission. However, being in the right area a week prior, and coming so close to a recovery, and now to accept the far-out position was a huge letdown, both morally and spiritually. But, we still had a mission to do and we were on our way. Our flight out would be more than two hours.

The flight out would be much further and, of course, much longer than last week's flight, matter of fact, approximately 400 miles further south from the planned primary recovery area. These many extra miles and minutes provided us plenty of time to check and double-check our recovery equipment and gear. This also gave us extra time to think about a lot of things and events that had happened in past missions and in our everyday lives, and to wonder what the future had in store for us.

A1C John Lansberry (squatting) and another loadmaster untangling the recovery rope and hooks from a parachute inside a C-119J after an aerial recovery in 1960. /Photo credit: USAF

The Discoverer 14 Agena spacecraft at Vandenberg AFB in 1960./Photo credit: USAF

Most of our aircrew had been together since the 6593d had been formed as a new, special research and development squadron back at Edwards AFB in August 1958. Our aircrew originally consisted of nine people. The addition of a tenth airman was deemed necessary so each aircrew could have its own aerial photographer onboard to film the event when one of our aircrews was lucky enough to successfully make the first midair recovery of a space capsule. Our aircrew was commanded by Capt Harold E. Mitchell from Greenfield, Illinois; copiloted by Capt Richmond Apaka from Kailua, Hawaii; 1st Lt Robert Counts, our navigator, came from Bellflower, California; flight engineer, SSgt Arthur Hurst was from Tazewell, Tennessee; our winch operator, TSgt Louis Bannick was from Hermiston, Oregon; the loadmasters consisted of SSgt Algaene Harmon from Randolph, Alabama; A1C George Donahou from Russellville, Arkansas; A2C Lester Beale from Portland, Maine; myself, A2C Daniel Hill, I hail from Chambersburg, Pennsylvania; and our aerial photographer was SSgt Wendell King from Stearns, Kentucky.

On every mission, especially the live Discoverer capsule recovery ones, I would think of my parents, William and Nellie Hill, back home in our small village of Edenville, near Chambersburg, Pennsylvania. Once the Discoverer program and the space capsule midair recovery attempts by our 6593d were announced by the Air Force to the press, by way of news releases, my parents would always be anxious to receive word of each Discoverer launch. So, every time we were deployed for a Discoverer recovery attempt, I knew that my parents and other family members would be back home rooting for us. My parents were very proud of what I was contributing to the Air Force and the space program.

My father, William "Bill" F. Hill, was proud for another personal reason. He was working as a sheet metal fabricator for the Fairchild Aircraft Company in Hagerstown, Maryland. He assisted in manufacturing the C-123, twin-engine and high wing, cargo planes that I had previously flown on at Sewart AFB, when I flew as a loadmaster with the 2d Aerial Port Squadron in TAC.

I often thought about the day that I surprised him when I walked into his plant work area at Hagerstown. I was the loadmaster on a flight that came to Fairchild on that day to pick up several aircraft parts for our Tactical Wing at Sewart. It was wintertime 1957, and I can remember the surprised look on his face when I showed up wearing my heavy blue material flight suit and shining aircrew member wings. His manager, my pilot, and copilot, along with myself, had set up the surprise. Due to the circumstances, they all allowed my father and me to have a fairly long visitation together. It was a great day for us and also for my mother when she received the word. Flying as an aircrew member in the C-119J aircraft was no exception, my father's hands helped to manufacture these airplanes as well. So, every mission that I flew on, either practice or live Discoverer recovery attempts, I always felt and knew that my father and my mother, with their close ties to Fairchild Aircraft, were a part of what I was doing in the Air Force. Fairchild, Hagerstown was the only plant that manufactured both the C-123 and C-119 aircraft. Being over 5,000 miles away from home, it was a warm comforting feeling.

> *"So, every mission that I flew on, either practice or live Discoverer recovery attempts, I always felt and knew that my father and my mother, with their close ties to Fairchild Aircraft, were a part of what I was doing in the Air Force."*

Long flights, as previously indicated, gave the recovery aircrews much time to ponder and think about various situations. Our crew had taken a good ribbing about not being able to get to Discoverer 13, since it did come down in our primary recovery area. It was a hard pill to swallow and it was difficult to be

The Pelican 9 crew and its C-119J that recovered Discoverer 14 on August 19, 1960./Photo credit: Daniel Hill and USAF

The Thor Agena with Discoverer 14 being launched from Vandenberg AFB on August 18, 1960./Photo credit: USAF

heading far south of every other recovery crew flying out that day. It was a feeling of, almost like sitting on the bench while the rest of the team participated in the play. Except, that it was not exactly the true picture, since we did after all, have a recovery area that needed to be covered. I guess it was a feeling that we were just in it for the ride. Let's just say, it was an uncomfortable position to be in.

Our aircraft and aircrew came very close to not even being part of the Discoverer 14 space capsule recovery attempt, due to problems with our aircraft. It was discovered on August 18, the same day as the Discoverer 14 liftoff, that our aircraft #037 had a leaking cylinder intake in our number one engine. After a serious, but quick, evaluation it was determined that the defective cylinder had to be changed (this was decided around the time that Discoverer 14 obtained its orbit). This meant that absolutely no time could be spared, so the maintenance crew, along with Sgt Hurst and Sgt Bannick worked long hours (some of the maintenance crew worked all through the night and into the early morning hours) to assure that our aircraft would be repaired, have its preflight, and be ready for our take off by mission deployment time.

Lou Bannick, our winch operator, who was mentioned early on about the subject of instigating to get something or someone stirred up, was in his usual form. As Capt Mitchell came into base operations on the morning of August 19, Lou approached and teasingly advised "Mitch" that it sure looked as if we would have to sit out this mission on the ground, because during the night they had a heck of a time with the cylinder repair, so the "Old Lady" (#037) would not be able to fly. However, Capt Mitchell knew Lou too well. Bannick was already wearing his flight suit, and that was a dead giveaway that the engine had been fixed and the plane was ready to go. Capt Mitchell advised Lou that it was still OK, and that he would fly the mission on just one engine, and all that Lou had to do was to assure that he and the rest of our aircrew be at the aircraft in advance (approximately forty-five minutes) of the scheduled take off time. Of course, Lou's insides had to be tickling as he felt satisfied with his acting and happily strolled off to round up the rest of the winch operators to match the coins to see who would pay for the next round of coffees.

On any long mission, providing that our job duties were up to snuff, it was permissible during the flight for us rear end recovery crews to play cards, smoke, chat or even take a nap. (Try sleeping in a very noisy C-119. It's a challenge, but it can be done if you are perhaps tired enough, and you convince yourself that a Boxcar built by the lowest bidder doesn't vibrate you apart.) Many a time, mostly on the way home after long hours out there, the recovery crewmembers would lay on top of the plywood covering that housed the huge extra fuel supply tank. I think that I had mentioned that the tank sat between the winch operator's station and the forward bulkhead prior to going up into the cockpit. Our "pillows" consisted of extra packed parachutes that were mandatory to be onboard. Also, on any over-the-water flights it was mandatory that not just the rear recovery crews, but all the crewmembers had to wear a "Mae West" floatation vest that could be inflated, if necessary. These were worn over our flight suits and then the personal parachutes were worn over the vest in case of emergencies. After all, our recovery duties working at the open rear end were and could be extremely hazardous at all times. Plus, the entire aircrew was at full risk for just being in a two-engine "obsolete" aircraft, flying over a "four-engine, modern aircraft ocean."

There were some trade-offs that helped ease our tensions about being out there on a limb, not to mention needing the dog leash straps that held us in the aircraft, the chutes, Mae Wests, and inflatable life rafts. The real trade-offs were the opportunities of just being out there, and also being part of the recovery business of the Discoverer program. Other pay-offs were the ability to fly and just observe what may be available on a given day, which inclued beautiful sunrises and sunsets, cloud formations of various magnitudes, the color variations of the different depths of water along and around the shores and reefs, sailboat races, smoke and lava from an occasional volcano eruption, whales on migration (on several sightings, a pure white whale), and the list could go on and on to mention the beauty of flying over and

Loadmaster SSgt Thomas Phillips wearing his floatation vest in 1960./Photo credit: USAF

The first recovery C-130 parked beside several C-119Js of the 6593d at Hickam AFB in 1960. A2C Daniel Hill's C-119J is parked beside the C-130A./Photo credit: USAF

viewing the entire Hawaiian Island chain. People pay thousands and thousands of dollars just to see some parts of the islands, and to think, we were getting paid to see all of this beauty out there, time and time again.

> *"People pay thousands and thousands of dollars just to see some parts of the islands, and to think, we were getting paid to see all of this beauty out there, time and time again."*

Nine aircraft from the 6593d, a C-130 assigned to the 6594th Wing, other support aircraft, both Air Force and Navy, along with several helicopter-equipped Navy recovery ships were all emerging on a direct course for each of their own recovery destinations. Based on the travel distance to get there, their times of departures had been staggered to accommodate the planned arrival times. Again, on this day our aircrew of Pelican 9 had the furthest distance to travel. The total Discoverer program's planned recovery area covered 36,000 square miles. The ballpark consisted of 60 by 200 miles (north/south—12,000 square miles) and part of its area was the primary recovery zone (the aircraft assigned here would be Pelican 1, five other C-119s and the C-130). The remaining larger recovery area (known as the outfield) consisted of 60 by 400 miles (north/south—24,000 square miles) that was to the south. Of course, the very far south end of this recovery area was where our Pelican 9 aircraft and crew were assigned. To the north of us in the outfield, two more C-119s were also assigned. This meant that our southern assigned aircrews had to cover an average of 8,000 square miles each. Let's say, we were far out, and had little, if any hope or chance of getting a signal or a visual sighting of the Discoverer space capsule.

We arrived at our designated recovery area at around 11:18 a.m., just a little over two hours from our 9:10 a.m. departure from Honolulu. Discoverer 14 was in its sixteenth orbit (approximately ninety minutes per orbit). We were just one pass (orbit) from the ejection and hopefully a capsule reentry, but not way down range in our way-out recovery position. As we were in it for the ride, the minutes ticked by, double-checks were performed again with our recovery equipment and gear. Tensions and anxieties were building, not with just our aircrew, but I'm sure, with all who were involved with the Discoverer program, especially with Pelican 1, I thought.

We loadmasters in the rear of the aircraft started the preliminary preparations during this waiting time. Due to some design changes to the hydraulic actuators, our initial recovery gear preparations were made a lot easier, faster, and safer. Sometime during our many practice missions (not sure when), the actuators that housed our recovery booms (poles) had been modified so that the top half could be hinged open and then could be closed and securely pinned after the poles and part of the harness were laid in place. In other words, we could now keep the rear doors of the aircraft closed while the first steps of gear prep and rigging were conducted. Looking back at the early days of the practice recovery missions over the Mojave Desert, I have already given some specific details about our rigging techniques. I explained how we then had to sit on top of the hydraulic actuators, dog-leashed to the inside of our aircraft fuselage and with the rear of the aircraft open, and reach outside to perform the cold masking tape and clip pin routine to secure the recovery hooks to the poles. With all that said, the latest modification with the hinged actuators eliminated the reaching out and straining against our dog leashes during the first part of rigging (amazing what new technology will do). Anyway, in the comforts of our Boxcar, with the doors closed, we had completed this first phase of the recovery gear preparation of hooks to poles. The placement of them into the open-hinged actuators would only come, if by chance, we would have any need to open the rear of the aircraft for a recovery attempt. Again, that was not likely; our being positioned at the very bottom of the recovery perimeter was "way out in left field," so to say.

Time almost seemed to stop as we continued the waiting game. Upstairs in the huge cockpit area, our flight engineer, Arthur Hurst, kept observing the aircraft gauges indicating engine rpm, oil and hydraulic temperature and pressures, assuring that all was going well with their performance. Navigator, Bob Counts was checking and verifying our precise location and adjusting the radio frequencies that would enable our aircraft receiver to respond to any space capsule beacon signal. As Capt Harold Mitchell and Capt Rick Apaka flew us around, they were also busy going over checklists and operational orders in the event Bob should pick up a signal.

Downstairs in the recovery area, winch operator, Lou Bannick was closely checking and going over the prescribed winch brake settings and double-checking the tightness of the nylon rope on the recovery winch drum. Loadmasters Al Harmon, George Donahou, Les Beale, and I were again checking and inspecting our recovery gear, using our standard checklist. "Did this harness look right? Did that harness pass inspection? Did this hook attachment look good? Did that hook setup meet approval? Was the 185 feet of half-inch nylon rope properly laced back and forth in the G-force reduction trough? Was it secured intact with the many 100-pound test breakaway cords that would spread the G-forces at the time of recovery? Should there be one? By searching and feeling, were any sharp areas found that could possibly cut and lead to line severing? Were the clevises still smooth and in place to assure a quick winch line to harness connection?" Our aerial photographer, Wendell King was delicately handling both of his cameras, a motion film and a still one. His checks were assuring a clean lens, power, and good lighting in case of a live Discoverer midair recovery.

Thoughts of last week's mission with the Discoverer 13 letdown kept coming into play, racing in and out of my mind. I remember wondering if our other crewmembers on Pelican 9 were going through the same thought process. We sure got a lot of ribbing from the other instigating aircrews. Matter of fact, we shared a full week of agitation generated by our friendly neighborhood capsule catchers. However, if the shoe was on the other foot, our aircrew would be doing our fair share of heckling. And, knowing Lou Bannick and his sense of humor as we did, he would be leader of the pack.

Discoverer 14 by now was well into its seventeenth orbit. The waiting game was getting shorter, but time didn't seem like it. The aircraft and ships at sea had earlier received communications that orbit 16 went fairly well. Should orbit 17 continue and receive good messages (as number 16 did) then there was good speculation that separation and ejection was possible as it passes over the Kodiak Tracking Station in Alaska.

Capt Mitchell had earlier come down to our recovery area for a short visit to see and feel reassured that our preliminary rigging and re-inspections of the equipment and gear were on track. He was a firm aircraft commander, but treated his aircrew members fairly and quite well. I was privileged to have the opportunity to serve with him. After all, we were and demonstrated as a good team. As indicated earlier, most of us had been together with aircraft #037 since the squadron startup at Edwards. Assured with our results, he climbed back up into the cockpit and assumed the aircraft controls from copilot, Capt Apaka.

As our Pelican 9 aircraft continued to noisily lumber over the Pacific Ocean using up our fuel and oil, time dragged on. We had discussed about how minutes seemed to grow longer as we sliced through the air. Speaking of oil, our squadron of C-119s was known to guzzle oil. I can't tell you how many times, but it was most of the time, that when we would look out the side fuselage windows, some signs of oil were visible. Behind the four-blade, propeller-driven engines, there was almost always a dark steady, heavy oil residue trailing along the sheet metal covering on the boom assemblies that held the tail and rear stabilizers to the wing assemblies. Sometimes the trail of this excess became thick enough to accumulate droplets of oil that would then run all the way to the rear vertical stabilizer. The wash from the propellers, along with the airspeed, would cause these oil droplets to blow off of the tail sections. More excess would continue to appear behind the engines, and the cycle and the process of losing oil persisted. However, anyone flying in these C-119s knew that basically the engines were good and trustworthy. If you didn't see excess oil, then it was time to maybe worry.

Finally, we heard the rpm of the engines increase and felt our aircraft start to climb to higher altitudes. Capt Mitchell had advised earlier, that his plan was to start to climb up to 16,000 feet approximately twenty minutes prior to the scheduled capsule ejection. This altitude would be 2,000 feet above our normal starting recovery level. He thought that performing this maneuver would give us a kick in the pants to assure extra airspeed, if necessary to advance to our target area as fast as we could, to safely get there.

During the climb, Lt Counts checked our position again with his loran equipment and got ready his beacon receiving system. We, down in the recovery area, were moving about, so we put on our oxygen masks and did some more checks throughout the back end of the aircraft. The plane eventually leveled and quieted down a bit, indicating that we had reached Mitchell's objective of 16,000 feet. Communications were silent while everyone waited for some indication of capsule ejection. Just prior to this separation, as the Agena A would come across a certain location in Alaska, gas jets would pitch the backwards orbiting spacecraft in a nose down, 60-degree angle.

As Capt Mitchell and Capt Apaka steered us through the skies, they were flying a pattern to keep us in our general recovery area. Time was moving on toward the space vehicle's separation. The schedule for this activation was planned to take place at 12:46 p.m. At this time, explosive bolts and springs would hopefully separate the capsule. If this phase went well, then a small retro-rocket would fire to slow the recovery speed of the capsule to less than 600 miles per hour. Then, after the parachute opening, a radio beacon on the capsule would start to transmit signals. Mitch and Rick continued the flight pattern for what seemed like endless minutes.

Finally, at least for this day, 12:46 p.m. was now history. Thoughts of what was happening up there were running rampant through, I'm sure, thousands and thousands of minds. I knew what had to be going through the minds of Pelican 1; they were hundreds of miles to our north in the prime recovery area, where we had the privilege to be a week ago with Discoverer 13.

Radio silence continued as everyone tried to stay calm; it was not an easy task. No word, not one word was uttered from anywhere. No word came from the Recovery Control Center who should be broadcasting that separation and ejection had occurred—silence, nothing.

Bob Counts remained glued to his navigational and radio equipment, his eyes strained into his oscilloscope, moving the frequency dial in search of a beacon signal. At 12:53 p.m., Bob broke the silence, "I think I've got a signal!!" He fine-tuned his receiver and both he and Mitchell immediately started to pick up a steady audio-beacon tone. Bob scribbled the frequency on his log and handed it to Rick Apaka to call the control aircraft. The cockpit was filled with anticipation. Bob had a good visual reading and both he and Mitchell started getting stronger audio signals. Downstairs we were listening on the intercom and naturally were anxious to see what lay ahead. Bannick sent us to our recovery stations and we all waited to see what was next.

The plane continued on the requested 360-degree turn. When doing this, Bob could home-in on the direction of the capsule beacon. Halfway through the second turn, Bob gave Mitch a heading to fly, 255 degrees. Our aircraft and crew started moving west by southwest and picking up speed.

Suddenly, Bannick motioned for me to follow. As we both scrambled to the forward bulkhead and up into the huge cockpit area, I was advised that he and I were requested by Mitchell to assist Sgt Hurst, and Capt Mitchell and Capt Apaka to search for our elusive target. Lt Counts would stay, eyes fixed into his equipment. Twisting necks, rotating heads, and straining eyes, and all that any of us could see was miles of sky and clouds.

With all the excitement and higher altitude, our entire aircrew was necessarily still on the oxygen mask system. On the intercom, Bob said that he was almost certain that the signal was straight ahead. He requested another 360 to make certain. Bob watched the signals on his oscilloscope change as we executed the turn. Following his strongest signal, he had Mitchell roll out of the turn. We were on our original bearing of 225 degrees. (We did not have any control aircraft telling us where to go this time.) Bob at that point stated, "I'll stake my career that it is directly ahead." Signals were getting more sensitive, an additional indication that we were closing in on our goal. Searching the vast sky around and above us, five faces, with oxygen masks, pressed against the cockpit glass. It had to be out there, somewhere.

Through his oxygen mask, came an excited shout from Capt Mitchell, "There it is! There it is!" Looking in the direction that Mitch was pointing, what the six of us in the cockpit saw through the windows that day was a most beautiful sight. Ahead of us and up 4,000 feet higher we made out an orange tint up against the background of towering cumulus clouds. What a sight to experience. It was 1:05 p.m., just nineteen minutes after the satellite had been reported orbiting over Alaska. In the twelve minutes that we had been tracking our invisible target we had flown 30 miles and descended 4,000 feet.

Following a brief observation of the descending chute, Bannick and I rushed down to our recovery stations and joined the others. Capt Mitchell then advised Lou to wait until he had reduced airspeed before opening the rear doors. Lou, Al, George, Les and I anxiously awaited his command. Wendell, with his cameras and filming equipment was ready. By the time our aircraft had slowed to 150 miles per hour, the capsule and chute had now descended to near our altitude.

Soon, Bannick received the word to open the rear doors and prepare for pickup. Assuring that our safety dog leashes were securely hooked to our parachutes and the inside of the aircraft, we loadmasters continued our recovery setup procedures. Harmon and I worked the right side, Donahou and Beale the left. As the doors were opening, each two-man team, handling our poles, placed the most outer ends with the two hooks and harnesses already attached into the cradles of the open hydraulic actuators. Adjusting the two recovery hooks and harnesses to the outmost part of the actuators, Harmon and Donahou then closed the upper-hinged actuator to the lower and secured the fastening pins.

At this point Beale and I handled the poles while Al and George handled the center harnesses assemblies with those hooks attached. As Les and I continued, we eased the poles to the outside of the aircraft while Al and George carefully allowed the harness assemblies to slide through their gloved hands and fingers until the poles and harnesses were in full trail behind our aircraft. Assuring that the poles were properly seated into the actuators, Harmon then made sure that the metal clevis that attached the winch line to the outside recovery harnesses was to the outside of the safety line-cutter assembly. Confirming that, he then seated the line and closed and pinned the assembly together. With all of the outside recovery gear now beyond the rear of the aircraft, the hydraulic actuators were operated to angle the poles down to a 45-degree position. We restlessly waited. Due to many practice missions, the time to open the doors, extend the recovery gear, and angle the poles took less than a minute.

Mitchell was near to completing a 360-degree turn under the parachute and preparing for a pass as the chute came through our altitude at 12,000 feet. On the intercom, he asked Bannick if we were ready. Bannick replied, "Roger" and ran through his recovery checklist: poles in pickup position, brake setting at 3.2, cable-cutter off. "Let her go," Bannick called, "but for gosh sake, don't invert it."

When Mitchell leveled out of our turn, we were about a 1,000 yards away from the capsule and chute. He had to maintain a 1,600-foot per minute descent at all times to keep the same altitude as the falling space objects. Lined up on target and closing in, at ten seconds out, Lou was advised to turn on the cable/line-cutter in the event of an emergency.

Harmon had the rearmost intercoms straddled over his head and we were positioned on the right rear of the open aircraft. He was ready to relay our observations of misses or hits. It felt good and refreshing to be standing at the rear as we were preparing ourselves for what lay ahead. Every time that we opened up the rear doors, just standing there seeing the new views and breathing in the fresher cool air was always an upbeat exciting feeling. Well, almost every time. The first time at Edwards, I was near the rear door when it opened. Seeing the back end open up, it appeared that the sky had opened up in front of me, with a scattering of clouds straight out in view and also below. While being hit with the cold air absorbing me, it was then that I wondered what I had gotten myself into. The first time out over Rogers Dry Lake and then further out over the Mojave Desert was something that I have always remembered. How forbidden and exciting was that first introduction to an open door practice recovery attempt that autumn day in 1958. But by now, nearly two years later, our aircrew had many miles of sky and practice recoveries behind us. Based on these past missions, we were professional capsule catchers, so we were told. Would the minutes ahead either deny or confirm this?

The feel of the aircraft at Capt Mitchell's approach seemed steady and smooth. No adjustment in flight was noticed, other than the descending force. We had to be close to the target. Uttering a quick prayer with open eyes, the contact time had to be about now. The huge chute had just come under the belly of our #037 and we flew just over the top of the brilliantly-colored chute, the capsule swinging slightly and both became gradually smaller as we continued on course.

Harmon reporting the near miss to the rest of our aircrew indicated that the chute had just missed our right recovery pole and hook by a mere 6 inches. Also observing this near miss, I knew that the chute with capsule had to have slipped to the right near the approach; Mitchell was usually true down the middle. We all breathed relief that we did not just brush the top of the parachute. Had that happened, in most cases, our recovery gear would have torn free of the poles. Attempting to re-rig in time would have been a minimal option.

Bannick switched off the cable-cutter. We flew the standard course, out twenty seconds as Capt Apaka counted off the time. On notice, Capt Mitchell maneuvered for a sharp 180-degree, always descending turn. With the half-circle executed he rolled out, lining up on the also descending chute and capsule. He stayed inbound for our second midair recovery attempt. Closing in again and ten seconds out, Lou again armed the cable-cutter. More quick prayers, the plane's attitude was smooth as we approached. Mitch must be down the middle, I thought. I watched the chute and capsule below drift underneath us again. This time Harmon's message to the rest up front was, "Two or 3 feet too high, captain, but you were right down the middle." Lou was advised once more by Capt Apaka to disable the cable line-cutter. After what had just happened on our second pass, I'm sure that "Aw, (*curse word*)" was on all of our minds in Pelican 9.

At least our aircrew of "Ole #037" was the only recovery personnel who knew exactly what was really going on with the status of Discoverer 14. The main reason for our big secret was due to the fact that when Capt Mitchell attempted to communicate with recovery command and advise them about our signal, search, and sighting of Discoverer 14, he was abruptly told to stay off the air and to maintain radio silence, so he would not "interrupt the recovery operations." After what I have found out in recent years, I would not be surprised if it was Col Moose Mathison that so hastily broadcasted that unexpected response.

Watching out the rear of the aircraft, we saw the capsule slightly swaying back and forth. We continued on Mitchell's descending flight path for another scheduled twenty seconds, and we saw the same thing that Mitch and the cockpit crew observed. Our aircraft was descending closer toward an overcast sky that lay below us at around 7,500 feet. We had just dropped below the 10,000-foot level, and would soon execute the 180-degree steep turn around, to again line up on the payload. We had now flown low enough to be oxygen mask free, so our energy levels were normal again as we continued our descent.

Mitchell and our entire crew were becoming more uneasy, and we knew full well that this attempt in motion would most likely be our last chance of making a historical midair recovery of the Discoverer 14 space capsule.

Capt Mitchell, remaining calm as possible under the circumstances, called Bannick, "Maybe we'd better actuate the poles to the full down position." (This meant from 45 to 60 degrees.) It was a chance, but a better one, to give us a larger space between the aircraft bottom and the recovery hooks. It was more risky to do this, because it would set up a chance of inverting the chute at contact. If this happened, it was a known fact from our many practice missions, that the snagged payload, recovery hooks, and harnesses could be torn from the winch line and lost to the ocean. It was a very tough decision that added more stress onto the commander, but it would be more stressful to not catch the capsule on our very

The parachute passes beneath the recovery loops during the second attempt to recover Discoverer 14./Photo credit: USAF

last chance and final attempt. Pressures mounted, especially on Mitch. Bannick with his own attempt at assuring Capt Mitchell piped up on the intercom, "Take it easy, sir. We'll get her this time."

Mitchell decided on the way out to make a tighter (shorter) pattern. Descending out of his final leg of the turn as the aircraft leveled out, this time he had put us only 500 yards from the capsule and chute. We were closing in and praying. Bannick armed the cable-cutter. Up front, Mitchell and the rest saw the canopy of the chute bobbing slightly and moving a bit to the left. He slipped the aircraft, easy, easy. A few seconds out he was right on line, but just prior to the chute going underneath the nose of the aircraft, the chute veered to the right. He wondered if he had missed our last chance!

I was standing near Harmon on the recovery floor near the opening to the clouds below us. I knew from past attempts from the familiar attitudes of the aircraft performance that the payload we were after was within feet of our recovery gear. Continuing to look out of the huge open space, I scanned the overcast of clouds that also now appeared to be not only below us, but seemed to be rising to our level. But, in reality it was Mitchell descending us closer and closer to their intensity. The capsule, chute and our Pelican 9 were in a simultaneous performance of dropping 1,600 feet per minute. "God, please let it happen. Please let it happen."

All together it seemed, but in a very rapid sequence of events, it happened! A visual of the chute canopy, right hook/pole contact, the feel of a tug on the aircraft, the flexing of the pole several yards more to rear and up, the hook/harness break away from the pole, the sound of the 100 psi cords exploding apart from the reduction trough assembly, Bannick's winch drum reeling out hundreds of feet of line, the rippling effect and sound of the chute as it collapsed into trail, and the, yes, the beautiful gold-plated space capsule as it swung into trail in a curved arch behind and a bit lower than the chute.

Suddenly everything seemed to stop except the sounds of our engines and airflow into the rear. Then a voice on the intercom, "A good hit, captain, on the right pole. We've got her in tow." It was Harmon's southern drawl. Bannick's fine experienced knowledge allowed the winch to payout around 350 feet of line (about half). The brilliant orange and silver parachute and gold-plated space capsule had come to us from space. It was ours to have, but it was not onboard the plane yet.

Apaka called to put the word out to the control aircraft advising that we had the capsule in tow and headed a course toward Hickam. Capt Mitchell came down to our recovery area to see his third-attempt catch, trailing behind our plane and just above the overcast of clouds. The catch took place at our 8,500-foot altitude. There would not have been enough clear sky for another recovery attempt. It definitely was our last chance for a midair pickup. "Thank God. He did let it happen."

Upon Mitchell's arrival downstairs he observed our conquering attitude. After we raised the actuators, we immediately retrieved the poles to the inside and stowed them. Then we took turns backslapping, handshaking, and discussing our proud and triumphant trip back to the States. Lou was standing over his winch running it at low speed, to keep it cool and not take any chances of possibly flaring up a malfunction that could cause us to lose our catch. By inches, feet, and yards, Lou reeled it closer. As it all neared the rear, he had to stop the winch so that Donahou could reach out and remove a grappling hook that was in the way near the cable-cutter assembly. With the reeling in process again, the chute and capsule was soon near the rear of our aircraft. As Beale and I held onto Donahou's and Harmon's parachute straps with one hand, they then leaned out straining against their dog leashes, pulled the chute in, and Les and I took possession. After securing the chute partially underneath us, the same procedures were used to reach out and grab the chute-to-capsule reinforced straps and lift the space capsule, so gently, onboard. The

The Pelican 9 loadmasters recovering Discoverer 14./Photo credit: USAF

capsule was now sitting on the aircraft floor between the four of us. Bannick switched the cable-cutter off for the last time that day.

The capsule was shaped like a large kettle drum, 33 inches across and 27 inches high. Having already cast off its retro-rocket and reentry heat shield, it weighed around 100 pounds. Its gold-plated sides gleamed brightly in the light streaming in through the recovery doors. We also took notice that the top was scorched and in some areas, handwriting was visible, probably from the launch preparers. The homing beacon was still broadcasting its signal and the strobe lights on top of the capsule were still blinking. Surprising to the rest of us, we soon discovered what Harmon had experienced earlier as he reached out and touched the capsule just prior to it being lifted onboard. The space capsule was still quite warm to the touch, resulting from the extreme high temperature that was absorbed by the heat shield during reentry. This was unexpected.

After some more brief celebrating and giving thanks, Capt Mitchell went back up to the flight controls so that Capt Apaka could come down to see our prized catch. After our entire crew had a chance to admire and feel our space capsule, we found the switch and turned the strobe lights off. Then we lifted the colorful parachute and capsule and carefully tucked them into a gray metal, cylindrical canister. The canisters were carried on each live Discoverer recovery mission, in the event that some aircrew would hit the jackpot. We did hit the top prize; the canister lid was then closed and padlocked by Bob Counts.

While the action of recovery attempts and the real recovery were taking place, Sgt King was recording on motion film and taking still camera shots, a lot of what he thought to be good coverage of this eventful day in history.

Our course was taking us back home to Hickam. We were wide-eyed and full of ourselves for what we had just accomplished. We were Pelican 9 for this day and proud of it. After all the ribbing that we had taken earlier concerning our position of being the least desirable recovery area, it had now turned into the most envious of all positions. As we continued our course, I kept thinking, perhaps all the crew was thinking the same thoughts, of our first days at Edwards, the many practice missions, all the Discoverer missions, and of course our loved ones that supported and did the rooting for our success. Thoughts back to a week prior with Discoverer 13 and now this Discoverer 14 capsule riding in the plane with us. I knew that all of us must have wondered what the Pelican 1, the prime recovery team, was thinking on their flight back home.

It was not long into our air travel to home base that I happened to be up in the cockpit discussing our good fortunes of the day. The frequency was dialed to pick up regular radio broadcasts. Soon we all heard the announcer break in with a news flash. That particular news flash was about us. He went on to say, "The Air Force today made the first catch of a Discoverer space capsule. Today at 1:13 p.m., a plane carrying a crew of ten, with the 6593d Test Squadron (Special) based out of Hickam AFB, Honolulu, Hawaii, made the midair recovery of a space capsule. The Discoverer 14 space capsule was launched yesterday, August 18, 1960. It was snagged out of the sky with a trapeze-type device trailing behind the aircraft. The capsule was caught approximately 385 miles southwest of Hawaii, at an altitude of 8,500 feet, on the aircrew's third attempt. The plane #037, code named Pelican 9, was piloted by Capt Harold E. Mitchell. Capt Mitchell and his aircrew of Pelican 9, today, made a historic first." With all that said about us, I knew that the word about us was traveling to the States and across the world as well. It would only be a short time until my parents and family in Pennsylvania would hear similar news break-ins. I also knew that #037, its aircrew, Discoverer 14, and yes, Pelican 9—our code name for the mission, would go into the history books and remain there.

The celebration for recovering Discoverer 14 started immediately upon their arrival to Hickam. The aircrew members of Pelican 9 were received with cheers as they opened the aircraft cargo door. The crew was awarded medals and then some of them were sent on a public relations' tour. Hill talks about their arrival at Hickam and the other events that followed.

We were the last recovery crew to land at Hickam. It was 3:35 p.m., hundreds and hundreds of people had gathered near our operations hangar to greet us as we taxied in. As soon as #037 came to rest in the chosen spot that Capt Mitchell was directed to park in, he immediately shutdown the engines. It was awfully quiet after flying the mission, but that sure changed as we opened both the rear and side doors to depart from the plane. We were met with thunderous cheers and applause. The wives of my teammates were there to greet them home and were a loving part of the ovation, as wives naturally should. Lights from the camera flashbulbs were vast.

I'm not sure who exited the front door first, but I remember that I was the first to jump down to the flight ramp out of the opened beavertail back door. I was then followed with the same 4-foot jump by Les Beale. We both turned around and while reaching up to the recovery deck, Harmon and Donahou handed us the padlocked container with Discoverer 14 safely inside. Waiting for all the other members of our aircrew to depart the plane and join us, Les and I then carefully carried our crew's recovered prize to a nearby table that was sitting on the flight apron, under the left wing of #037. Gently we placed the package on the tabletop. Again the huge crowd cheered and applauded, bulbs continued to flash. There were many news/press people on site.

Col Teuvo Ahola directed us to form a line so that we faced the table, which had now been moved several yards to allow more room for a microphone system. Our minds were swimming at this point and it's hard to remember all the exact details, but the colonel then introduced none other than Pacific Air

Chapter 2 - A2C Daniel R. Hill

Forces (PACAF) Commander, four-star Gen Emmett "Rosy" O'Donnell. The general gave a brief, but an appreciative and applauding speech directed to our efforts and the Discoverer program. At the end of his speech, and surprising all of us, he had awards that had just been approved by Gen Thomas White, Chief of Staff. He started with Capt Mitchell and presented Mitch with a Distinguished Flying Cross, then to the rest of our aircrew, he presented and pinned to our flight suits, an Air Medal for each, starting with Capt Apaka and ending with me. I was on the opposite end of the line from Capt Mitchell. I took claim to being the youngest and the smallest member of our now famous aircrew.

Time seemed to stop as applause and flashing lights continued. Then as the general stepped aside, the aircrew's wives rushed in with hugs, kisses and embraces. I remember stepping aside also; I have a film to verify that maneuver.

Also, during our welcoming a birthday cake was brought out and presented to me. Unbeknownst to Capt Mitchell and the rest of our aircrew, my twenty-first birthday had been the day before. "They put your birthday present in orbit yesterday, Danny, and it was delivered today," Col Ahola announced, then, more cheers, applause and flashes.

> *"'They put your birthday present in orbit yesterday, Danny, and it was delivered today,' Col Ahola announced, then, more cheers, applause and flashes."*

The container with Discoverer 14 inside was whisked away, placed inside a C-130 aircraft, and flown by some of my friends to Sunnyvale for studies and observations. After that it went on to Washington, D.C.

By the time that I called my parents back in Pennsylvania, it was late at night there. They were thrilled, as well, over our success that day. The call that I made to them was one of many that they had received since the news releases. They had been called and interviewed by numerous newspaper reporters and some TV folks. Some had been local and some were reporters from the big cities. There would be more the next day. I knew how proud they were.

In Hawaii, the celebrations and celebrating went on for many hours—enjoyment that was long overdue. Our aircrew got the attention for being at the right place at that right time.

Reviewing the events in my mind, especially the many aerial recovery missions that lead up to August 18 and 19, 1960, the Discoverer 14 launch, and then aerial recovery, I do believe that, #037, Pelican 9, and we were destined to "Catch a Falling Star," the first midair recovery of a satellite capsule from space. Our 6593d Test Squadron (Special), the Air Force, and the Discoverer program were victorious, finally.

The words that Col Ahola spoke to me in front of the mass of spectators and press, about my birthday and birthday present, kept running through my mind. I thought over and over, "God did provide me with a great birthday and present."

The media coverage started as soon as the Air Force released the word of our successful recovery of Discoverer 14. Of course, the news was traveling fast worldwide. Broadcasts on both radio and television, I am sure, were being sent out to listeners and viewers, similar to what we had heard on the aircraft radio on our return flight back to Hickam that mid-afternoon on August 19, 1960, with Discoverer 14 safely tucked inside our aircraft.

The arrival of Discoverer 14 (within the metal container) at Hickam AFB on August 19, 1960. Left to right: TSgt Louis Bannick, A2C Lester Beale, A1C George Donahou, SSgt Arthur Hurst, A2C Daniel Hill, SSgt Algaene Harmon, and SSgt Wendell King. /Photo credit: USAF

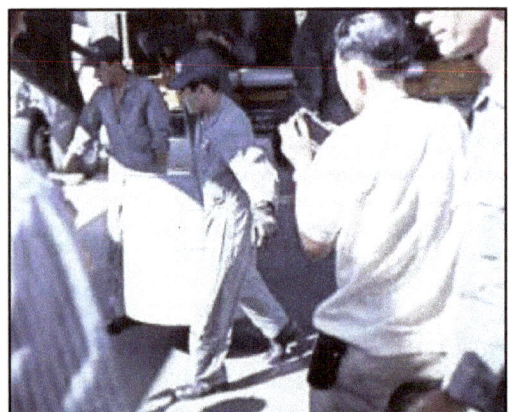

The Discoverer 14 container being removed from the C-119J at Hickam AFB. (Left) A2C Lester Beale and A2C Daniel Hill carrying the Discoverer 14 container. (Right) The Discoverer 14 container with A2C Daniel Hill behind it./Photo credit: USAF

Chapter 2 - A2C Daniel R. Hill

Then, after we taxied up to our designated spot at our hangar and opened our aircraft doors, the flashing of camera bulbs started the more personal media attention to both our aircrew and then individual members. We were interviewed or I should say, we were asked a few questions. There were a lot of press and some television reporters on hand. We had been briefed earlier, pertaining to the news media. There was little that we could say about matters of the recovery operation, our squadron, or the Air Force. When someone would start to ask questions, it was always an uncomfortable situation. We were always half afraid that we would say something that was out of line or context. We definitely did not want to upset our leaders or the Air Force.

My birthday was on August 18, 1960. The announcement of it by Col Gus Ahola on the following day, the day of our recovery of Discoverer 14, brought much attention from the media. The Associated Press (AP) picked up a photograph of me being presented a birthday cake, and it probably got some worldwide attention since it was in most of the large city newspapers in the States.

A nice lady who lived in Brooklyn, New York, sent my mother a copy of our recovery of Discoverer 14 and also a copy of the photograph of my birthday cake presentation. She told my mother that she had cut it out of the *New York Times* newspaper. She told my mother that she must be proud of me. The lady informed my mother that she had seen that I was from Chambersburg, Pennsylvania, so she called the Chambersburg Post Office to get my mother and father's rural address. Anyway, she just wanted my mother to know that the article along with my picture was in the *New York Times* and wanted her to have

Gen O'Donnell awarding the Air Medal to A2C Daniel Hill./Photo credit: USAF

it. As mentioned before, my parents got a lot of phone interviews with the news and television reporters, both locally and from big cities across Pennsylvania and Maryland. That made them, and me, feel good and also proud.

After some celebrating, Capt Mitchell, 1st Lt Counts, and TSgt Bannick left for the United States almost immediately and represented our entire aircrew with various types of media. The rest of our aircrew followed later, upon an invitation to Washington, D.C., by Lt Gen Bernard A. Schriever.

All in all, we got much attention and news coverage for that time in history. Along with it, in some corners of the world, news outlets gave my birthday interviews and photograph as much attention as the space capsule itself. That alone, made me feel famous and proud. In a way, when the attention was directed at me, because of my birthday and being the youngest of our recovery aircrew, I felt awful at times, that perhaps I was taking something away from the rest of our Pelican 9 crew. But, I also realized that I didn't cause the situation, it was just the circumstances of events that surrounded us at that time. I received at least fifteen birthday cakes from mainland bakeries (well wishers), and as the days continued there were more. So, I'm not sure of the total numbers, but beyond our needs, the balance was donated elsewhere.

The entire crew was honored at a formal dinner with several top generals. Hill notes they were treated like celebrities. They went to Sunnyvale and Vandenberg AFB to tour the satellite operations facility and the launch site. For Hill, it was an unbelievable experience.

On August 24, 1960, I received a personal invitation to attend a press meeting and luncheon at Andrews AFB, Washington, D.C. Gen B. A. Schriever, commander of the Air Research and Development Command, invited our entire aircrew to attend a press meeting at 11:00 a.m. and a luncheon at 12:00 p.m. on August 26, 1960. Already in the United States, performing a goodwill tour was Capt Mitchell, Lt Counts, and TSgt Bannick. The balance of our aircrew would meet them there.

We flew from Honolulu on a Pan American Airways commercial jet to Los Angeles and then on up to San Francisco. Everywhere we went, someone would introduce us as the famed Pelican 9 aircrew that had just made the midair catch of the Discoverer 14 space capsule. These introductions were at all of the airport terminals and to the rest of the passengers while we were in flight. Everywhere that we went, we were treated like real celebrities, with autograph requests, and we even turned down propositions from some admiring women that were advanced.

The reason for the San Francisco leg of our trip was an invitation to tour the Sunnyvale space operations and to meet the brains behind the space race endeavors during the Cold War years. By invitation, we also went to Vandenberg AFB to visit the launch facilities where all of our Discoverer rockets were counted down and lifted off. Again, everywhere that we went, we were met and treated with the utmost respect and recognized as celebrities, even by the top brains of the Discoverer space program, both military and civilian. I thought, as probably did the rest of our aircrew, that this couldn't be happening to me. Most of the time I felt out of place, but considering everything, I handled it all well. After all, we had proved ourselves professionals with the successful historic Discoverer 14 catch. The tours were very educational.

It was now August 26, 1960, when we arrived in Washington, D.C., one week to the day after our prized catch. After checking into our plush downtown hotel, we were then picked up by several of Gen Schriever's staff and escorted to the Air Research and Development Command at Andrews AFB. Upon arrival there, we were taken into a huge briefing room. There, in addition to seeing more of the general's staff, we were reunited with the other three members of our aircrew, Mitchell, Counts and Bannick.

Several of the Pelican 9 aircrew on their way to Andrews AFB to meet with Lt Gen Bernard Schriever in August 1960, left to right: Capt Richmond Apaka, A2C Lester Beale, A1C George Donahou, SSgt Algaene Harmon, A2C Daniel Hill, SSgt Arthur Hurst, and SSgt Wendell King./Photo credit: USAF

The Douglas Aircraft blockhouse crew and the Discoverer Control Center at Vandenberg AFB in 1960 posing with the Discoverer 14 capsule and parachute after its recovery from space./Photo credit: USAF

We were busy with introductions back and forth between us and the staff, plus we were introduced to three other visitors who were also invited there by Gen Schriever. The other three invitees were Col Alvan N. Moore, Director of Recovery Operations; Maj Robert M. White, X-15 test pilot from Edwards AFB; and Capt Joseph W. Kittinger Jr., experimental balloon/gondola parachutist from Holloman AFB. With all of the introductions and several more briefings out of the way, in walked a tall and young-looking general. With all six of his stars sparkling from the lights overhead, I knew it was the general who had invited us there; it was B. A. Schriever, "the missile man." Suddenly, you could have heard a pin drop. All staring eyes were fixed on the general as we all snapped to attention. Then we were put at ease.

Although it was unnecessary, the general was introduced by one of his staff officers. Gen Schriever then welcomed all of us to Andrews and to the press meeting. He continued with thanking us for coming to visit both him and the Air Research and Development Command Headquarters. Then as he reviewed each historical event that had taken place the week prior, he praised and also thanked us for our successful accomplishments. He went on to say that in recognition of the ARDC achievements that had been successfully performed August 9-19, 1960, the Air Force and the nation were equally proud of our historic significance, and that the Air Force wanted to show tribute to our performance. It would honor each one of us in a way that would be recognized in both the present time and for years to come. While holding up a couple pieces of paper, as the staff quickly handed us a copy, the general started to read. He was reading from a charter outline with his signature affixed near the bottom of the second page. The charter was written for a newly-formed organization. At that point, he further revealed an unincorporated association to be known as the "Aerospace Primus Club." The objectives and purposes of the club were: to promote and encourage original accomplishments in the discovery, development, testing, and use of equipment and techniques in furtherance of the aerospace program of the United States Air Force. Also to recognize and honor the first individuals to attain or participate in aerospace accomplishments deemed to be of historical significance.

At this time in the program, we were advised by the general that thirteen of us would become charter members of "The Most Exclusive Club on Earth," the Aerospace Primus Club. With all of that said, our hearts were really pounding and our heads were swelled with pride, at least mine was. Then, one by one the general called our names to approach him at the front podium. As I stood facing the general, he looked down into my eyes, thanked me personally, and while shaking my hand, presented a plaque to me that honored me into the Aerospace Primus Club.

As I returned to my place at the press meeting, it was then that I had a chance to look at and read the engravings on my newly acquired plaque. It read, from the top to bottom: "Most Exclusive Club on Earth—Honors—A2C Daniel R. Hill—AF 13606761—Loadmaster—Aerospace Pioneer Emeritus—First Aerial Recovery of Satellite from Space 19 August 1960—B A Schriever Commander ARDC 26 August 1960—Aerospace Primus."

Our entire aircrew got our own personalized plaque for the successful midair recovery of the Discoverer 14 space capsule. Also receiving a plaque was Col Moore for the superb planning, leadership, and directing of the recovery operations; Maj White for piloting the X-15 aircraft to a record altitude of 136,500 feet, flying higher and faster speeds than man had ever obtained; and Capt Kittinger for making an open gondola balloon ascent to the record altitude of 102,000 feet, more than 19 miles above the New Mexico desert. At this altitude, he jumped for a parachute descent by means of an experimental stabilization parachute system, free falling 16 miles for four minutes and thirty-eight seconds to an altitude of 17,500 feet where a deployment of the man-recovery parachute occurred.

Chapter 2 - A2C Daniel R. Hill

Lt Gen Schriever presenting the Aerospace Primus Club plaque to A2C Daniel Hill./Photo credit: USAF

A2C Daniel Hill displaying the Aerospace Primus Club plaque.
/Photo credit: USAF

What a day. We were all without our aircraft, but we were all on cloud nine. It kept running through my mind, "Man, how did I get mixed up with these guys? What did I do to deserve this?" I can't tell you just how very thrilled and proud that I was!

Nearing the end of Gen Schriever's special ceremony and presentations, we were reminded by him of the luncheon that followed. During this luncheon we were then invited to a formal dinner that would follow in the evening, where we, the thirteen inductees, would again be the guests of honor.

Now, here I am, a 21-year-old airman second class, still amazed at just being there, and to top it off, being a guest of honor was a little hard to believe. Especially, when I looked around the huge dining area and saw mostly top ranking officers and their wives. There were some important civilians there as well. In addition to Gen Schriever there were other generals present with names that I had heard of before, but never dreamed that I would ever have the chance to see, let alone meet. I think most of the top Air Force generals were there for this affair. Now, here I was, wining, dining and conversing with the elite and their wives. (I used to feel out of place at our squadron's beer busts, sometimes talking to lieutenants, captains, and occasionally our Maj Nellor.) There I was, and all of this brass and their lady folks were popping me questions left and right. For a young kid, originally from the orchard country of Pennsylvania, I thought that I did quite well. I was rather proud of myself.

We newly chartered club members were each assigned to sit at different tables, so this would spread us out among the crowd. I think it was Mrs. Schriever on my left and to my right sat a distinguished looking general. He was quite interesting and a very remarkable person to talk with. He was retired Maj Gen Frank O'Driscoll Hunter, a leading World War I fighter ace. We had some very good and interesting conversations back and forth about our Air Force experiences. Hunter AFB, Savanna, Georgia, was named in honor of this general. He was one of the few living individuals who ever had a base named after him. All in all, it was a great joint gathering of air and space personnel and it was an extremely exciting time in my life. It was an honor to be there, and it was especially an honor to be personally recognized by the top heroes and leaders of our United States Air Force.

Between the special arrangements that Gen Schriever had provided and our time to fly back to the West Coast, I had around a half a day of free time. A buddy of mine, who I had called, came to my hotel, picked me up and drove me to my parents' home in Edenville, Pennsylvania. My parents knew that I was in D.C., but didn't have any idea that I would make it home. It was a quick surprise visit; it had been around two years since my last trip home. We shared a good time together, even though it was only a couple of hours. I filled them in on the events from yesterday and then it was an hour and a half trip back to D.C.

On our way back to Hickam, we were taken to San Francisco. There we toured and were then the toast of the town at a variety of nightclubs. Everywhere that we went, we were still being highlighted and introduced as, "The famous Pelican 9 aircrew that made the first historic midair catch of a space capsule, Discoverer 14."

After a few stops, the Discoverer 14 capsule was returned to Hickam AFB. Hill was part of the capsule display tour that went around Hawaii telling the story of Discoverer 14. The capsule was later taken to the Air Force Academy. It is now a display at Wright-Patterson AFB in the National Museum of the US Air Force.

Sometime in September and on into October of 1960, several of us from our 6593d Test Squadron (Special) had been chosen to take our Discoverer 14 space capsule on display tours at various locations throughout the island of Oahu.

Chapter 2 - A2C Daniel R. Hill

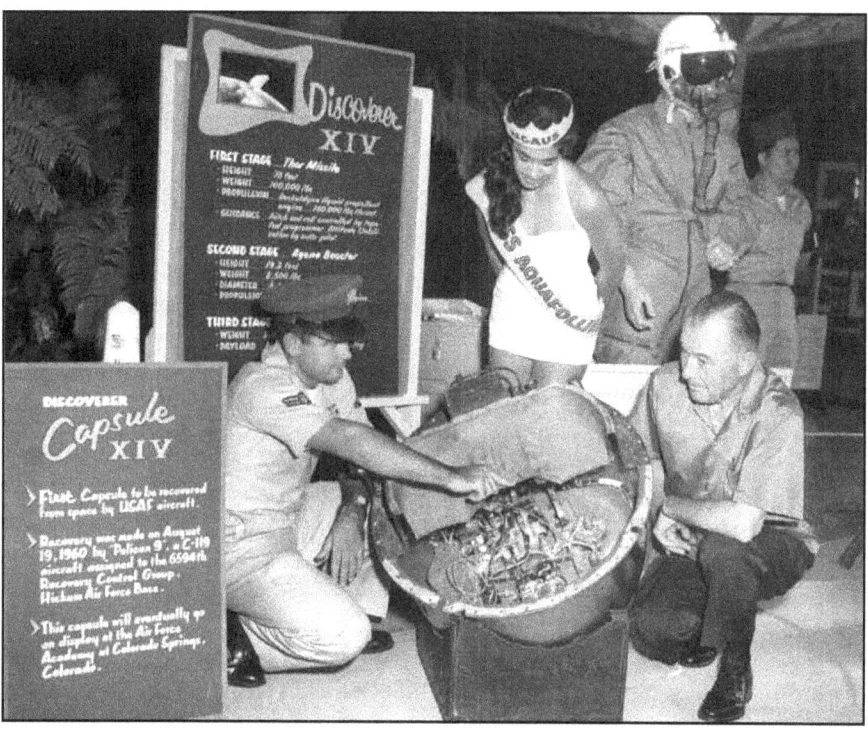

A2C Daniel Hill during the Discoverer 14 publicity tour./Photo credit: USAF

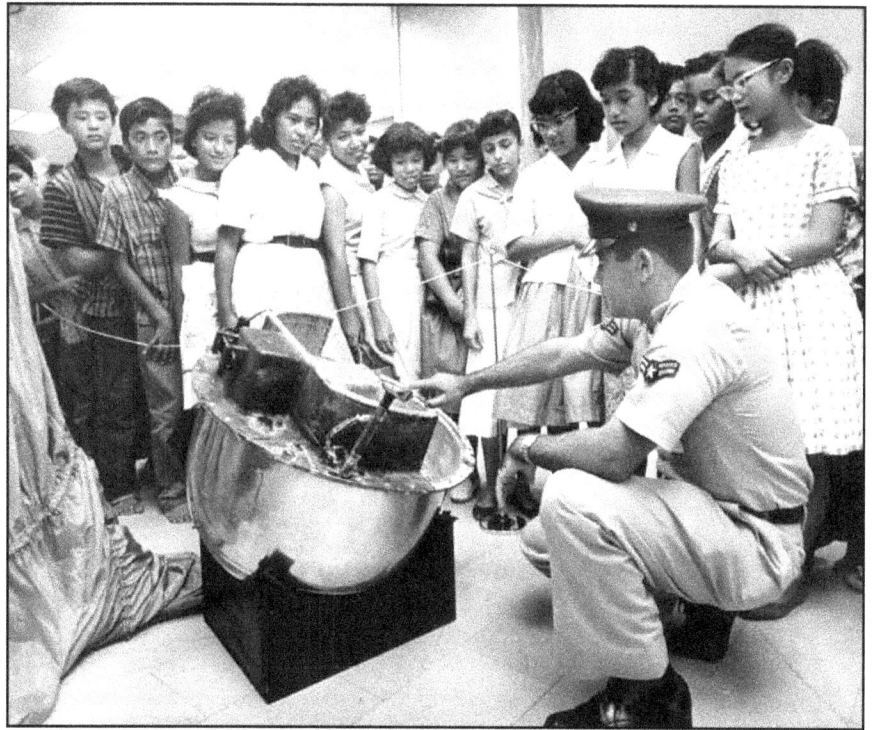

A2C Daniel Hill telling a group of children about the Discoverer 14 capsule./Photo credit: USAF

Immediately after the space capsule had been taken to Sunnyvale, it was stripped of all the scientific data that was pertinent during its historical liftoff, seventeen orbits, reentry, and midair recovery. It then went to a laboratory in Boston, Massachusetts, then on to Washington, D.C. About a month later it was returned to us at Hickam. It was still in the same gray metal, padlocked container that we had carefully tucked it into while on our Pelican 9 aircraft, back on August 19, 1960.

Several display panels with facts and statistical information were prepared describing the 450,000-mile journey of Discoverer 14. We would place these panels near the space capsule when we set up our tour displays about Discoverer 14 and our performance during its recovery. The 6594th Recovery Group provided us with the overall display; our squadron was under this group. The display only described the 6594th Recovery Group and not our 6593d Test Squadron (Special). However, we were in control of the display, the narration, and the performance program. So, right off the bat the local crowds knew full well that we were part of their hometown 6593d recovery team and part of the Pelican 9 recovery crew.

Only three people from our recovery squadron and our recovery group were selected to take the Discoverer 14 space capsule on the Hawaiian tour. The other members, other than me, were Capt Rick Apaka, our copilot, a native Hawaiian from the island of Kauai, and 1st Lt Bob Counts, our navigator. I accompanied either one of the officers, and sometimes I would be on the display tour on my own. We were always provided security personnel to guard our precious space capsule, the balance of the display, and of course, we tour masters.

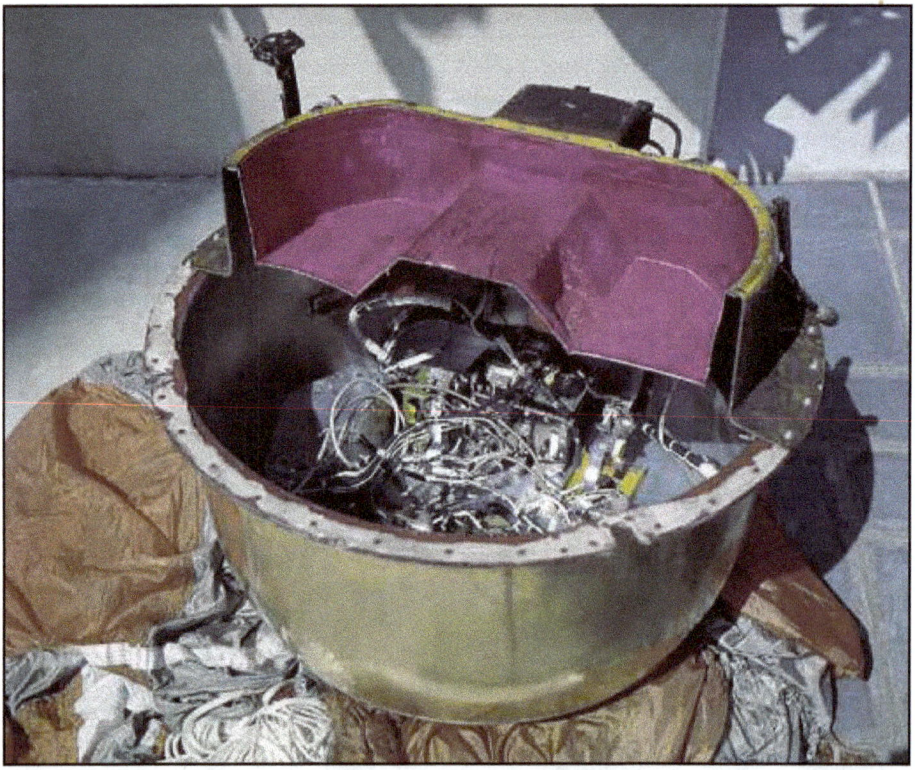

The Discoverer 14 capsule and parachute on September 7, 1960 at the Space Technology Laboratories' (STL) R&D Center (that later became Area A of Los Angeles AFB) in El Segundo, California. (Note how the capsule cover was cut in half.)/Photo provided by Northrop Grumman

Chapter 2 - A2C Daniel R. Hill

Daniel Hill with his former C-119J (#18037) at the National Museum of the USAF in 2008./Photo provided by Daniel Hill

I cannot remember exactly how many displays and narrations we conducted, but they were many and often. Our Hickam AFB Public Relations Office selected the areas where we performed. They then coordinated with the officials who were responsible for the specific locations where we provided the narrations. We toured and spoke at elementary and high schools, universities, various other military bases, some leading businesses and banks. I was generally noticed as the one who got the space capsule for a birthday present.

Overall, the tours went very well and we became professionals at doing them. It was just one more chapter in my young life that I was proud to be a part of the Discoverer program. I was also proud to have been the only enlisted airman to provide this additional service to the home public of Hawaii, representing our aircrew, squadron, group, Hickam AFB, the Discoverer program, and the United States Air Force.

I do not recall if the Discoverer 14 space capsule cover was returned to us with the capsule when we went on tour. Perhaps half of the cover remained stateside when they removed the film and canister. They may have taken the entire cover off when stateside, then cut it in half, then upon sending it back to us for the display tours, put only half the cover back on. With the cover half open, anyone would realize what the entire cover would have been like, plus it allowed everyone to look inside the capsule. I wish I knew more.

Upon completion of the Discoverer 14 Hawaiian tours, the space capsule was then scheduled to travel to the Air Force Academy at Colorado Springs, Colorado. I am not sure how long it remained there on display. The historic Discoverer 14 space capsule, along with its orange and silver parachute, and also its four-pronged bronze recovery hook are on permanent display at the Air Force Museum, Wright-Patterson

AFB, Dayton, Ohio. Please note that our Old Lady, our famous C-119J aircraft #037 was retired many years ago and is also on permanent display at the Air Force Museum along with Discoverer 14.

I have some friends who had been to the Air Force Museum and reported back to me that they had seen my #037 aircraft and my Discoverer 14 birthday space capsule. They informed me just how very nice the displays are at the museum and brought back some photographs for me. Back in 1985, when our family lived in Tennessee, I called the museum in reference to the display of #037. At that time, I was informed that the Old Lady was getting a "facelift" and that we may want to visit at a later date, because she would not be available for display for some time. Unfortunately, to this date we have not visited the museum. However, in the very near future, my wife Marilyn, perhaps other family members, and I are going to see my famed acquaintances at the museum; it is long overdue. Along with us, everyone should make an effort to visit these historic objects of the Discoverer/Corona program epoch.

Hill recounts fond memories of the "Old Lady," his C-119. At the time of his interview, he had not been to the National Museum of the US Air Force. It truly was a proud period of time in Hill's life.

Our winch operator, Lou Bannick, introduced the name "Old Lady" for #037. He used the name quite often. He and some of the other old timers in our squadron grew up with these C-119 planes, and knew them like family. They could fix them and they knew them like the back of their hands. I wish that Lou was still alive. He would have most certainly been a charm for you to interview. You would never forget it.

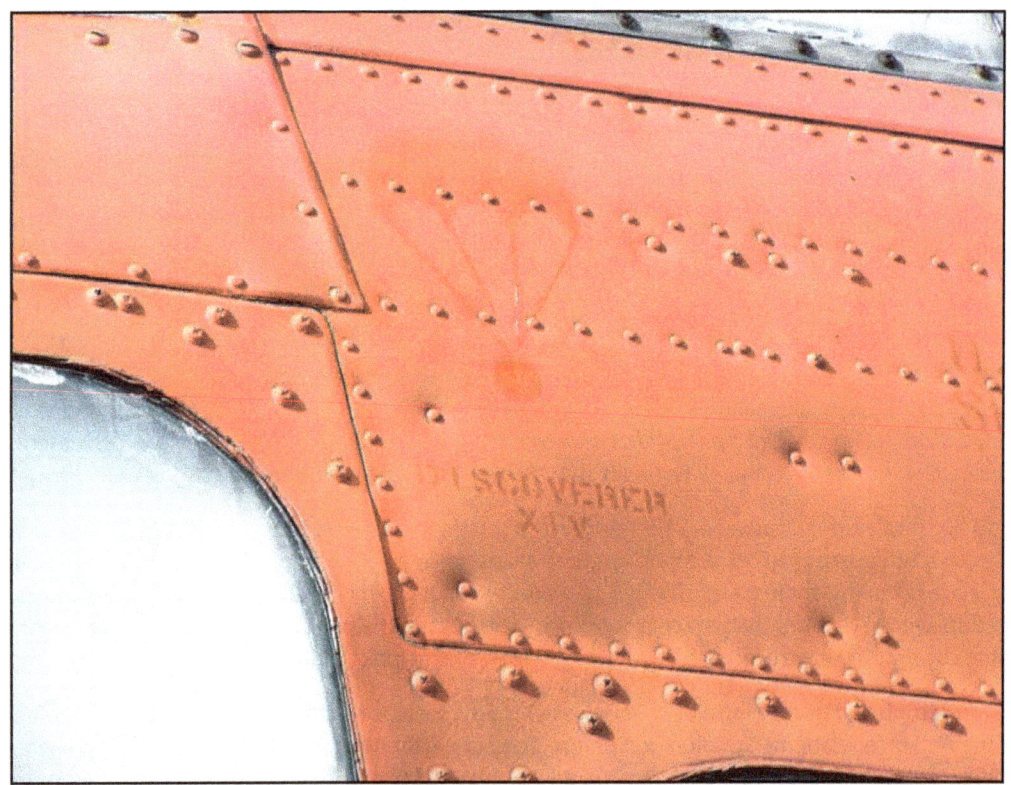

The fading parachute decal for recovering Discoverer 14 under the pilot's cockpit window of C-119J (#18037) in 2006. /Photo provided by Robert Mulcahy.

Chapter 2 - A2C Daniel R. Hill

One thing that impressed me, and I'm sure all the other aircrew members (especially Capt Mitchell), was when someone painted a parachute and capsule on the forward left side of our #037 aircraft. It was a status symbol that represented our catch of Discoverer 14 on August 19, 1960. At that time, it was the first and only one in the world. It felt great to see it there and of course we were all proud to be part of it.

Hill briefly discusses an issue regarding recovered Discoverer capsules, and security for the Discoverer program. For the crew it was mostly business as usual.

As far as our midair recovery end of the Discoverer program, I really don't know of any significant security or secrecy changes than we had before.

I will note that when we flew out hoping to recover Discoverer 13, I was not aware of a padlock for the gray metal capsule container. In my opinion, the reason for the padlock (if we were successful in the recovery, and we were with Discoverer 14) was because when the Discoverer 13 capsule was being transported in the C-130 aircraft from Hickam to Sunnyvale, it was reported that a certain high ranking staff officer (not in the 6593d) used his own authority to take the top off of the capsule and rummage through the contents that were inside of it. I was told about these incidents at the last 6593d Test Squadron (Special) reunion, and I have also recently read about it in Discoverer/Corona reference books. Anyway, the introduction of the padlock with the Discoverer 14 catch would hopefully keep unwanted fingers off of the capsule and its contents until it was in the hands of the folks back at Sunnyvale who had prepared the capsule and contents for its historic voyage.

But on the other hand, today, knowing that Discoverer 14 was the first space capsule to be recovered with Corona cameras and film onboard, I feel sure that security and secrecy were turned up a notch or two. But again, from our midair recovery end, things were about the same as far as I'm concerned. There were some things that we could not discuss about the program and we were definitely not aware of any "Corona" program. In those days, we only suspected that perhaps there could have been a reconnaissance camera onboard. After all, the news media had put those speculations out to the world audiences, before and especially after the Discoverer 14 catch. To us recovery aircrews, it was business as usual. Our job and mission was to continue to "Catch a Falling Star."

Hill describes missions after Discoverer 14. Hill's last Discoverer mission was 17. He did not reenlist. It was hard to leave, but Hill explains it was the best decision at the time.

Our aircrews of the 6593d Test Squadron (Special) were involved with all of the live Discoverer attempts. After the successful recovery of Discoverer 14, the next launch of the Discoverer program was on September 13, 1960, when Discoverer 15 made orbit. On reentry the retro-rocket's orientation was bad; the recovery vehicle overshot our recovery areas and splashed into the Pacific Ocean near Christmas Island. The capsule was later sighted, but unfortunately sank before it could be recovered. On October 26, 1960, Discoverer 16 was launched from Vandenberg. The Agena's "D" timer stopped during liftoff due to a power interruption, causing the Agena to fail in its separation from the Thor booster. Naturally, the orbit could not be obtained without separation.

Discoverer 17 was launched on November 12, 1960, and had a successful orbit. This was the first space orbit that used the Agena B. On the thirty-first orbit the recovery vehicle separated and the space capsule reentered the earth's atmosphere over our recovery zone. Our squadron's operations officer, Capt Gene W. Jones and his aircrew successfully made a midair recovery of Discoverer 17. Our aircrew in #037 was also at the scene, which at that time made us the backup recovery aircraft. If Capt Jones and his crew had missed their recovery attempt, Capt Mitchell and our aircrew would most likely have made our second midair recovery of a space capsule.

Capt Gene Jones recovered the Discoverer 17 space capsule on November 15, 1960. The C-119J seen in this photo was probably Daniel Hill's airplane./Photo credit: USAF

What another beautiful sight it was that day, to be flying and observing the capsule and chute in trail behind their aircraft. It was a good feeling to know once more that our squadron was ready and able. And, also to know overall, that the Discoverer program was getting back on track by placing the capsules in our recovery patterns. Our aircrew's involvement that particular day, and coming so close once more, brought back fond memories of our earlier successful Discoverer 14 midair catch.

Several months after our August 19, 1960 midair recovery of Discoverer 14, Capt Mitchell was flying us on a practice training flight when he was contacted that the Pacific Missile Range would like to have their nose cone back if we could catch it. The nose cone was a five-pound Arcas payload. Normally, the Arcas nose cones were not recoverable. But, the Pacific Missile Range decided that if the Air Force could catch it, they could then inspect it for heat damage and make a check of the instruments.

Capt Mitchell was given permission to try for the catch. So we proceeded to the area that we were directed, checked our recovery rigging, and anxiously waited. This would be a different catch than we were accustomed to performing. The earlier rocket was fired from the Pacific Missile Range launch site on the Island of Kauai (Capt Apaka's home island).

Chapter 2 - A2C Daniel R. Hill

This would be a rare catch, the first for an Arcas nose cone, which would have been on its 38 to 40-mile high ride into space and returning on a small descending parachute. The Arcas was launched at 10:54 a.m. We pressed our faces against the cockpit windows searching for a visual sighting of the parachute. After sighting and awaiting its descent to our altitude, Capt Mitchell flew by the small chute and extra small nose cone for a visual inspection. He proceeded on course for a number of seconds, made the 180-degree turn, and lined up on the 15-foot diameter parachute. Seconds later, perfect contact was made with our recovery gear. It was 1:30 p.m., and we had a rather small chute and a tiny nose cone in trail behind #037. The recovery was made around 2 miles high and 23 miles northwest of Kauai.

With the spur-of-the-moment aerial recovery safely onboard our famed Discoverer aircraft, we flew on home to Hickam. Of course, when we got back to base, a photograph was taken of the chute and nose cone. The entire story and photo were in the Honolulu newspaper the following day, indicating that our aircrew who made the "big catch" in the Pacific last August 19 when we caught a Discoverer nose cone in midair, made a "little catch" yesterday.

The last live Discoverer mission that I was involved in was Discoverer 17 in mid-November 1960, when we became the backup recovery plane. Based on the sequence of a monthly Discoverer liftoff schedule, I knew (or at least thought) that this would be my last 6593d Test Squadron (Special) live recovery mission. During that particular mission, from take off to landing, I breathed in and cherished every sight, sound, and motion that one could possibly savor that day. Although that was my last Discoverer mission, I still continued practice midair recoveries up until my last days with the famed squadron.

My enlistment time was completed during the fall of 1960, and I chose not to re-enlist. It was a very difficult decision that weighed on my mind for quite some time. I was basically frozen out of advancement for a long period of time. The reason for this was due to people with higher rank than me cross-training into the loadmaster jobs in order to get on flying status and also to receive the extra flight pay. We had higher grades transferring into my field from food service, supply, maintenance, and other modes of operation. Anyway, this froze me from advancing higher and also advancing faster, so at that particular time, although I did not want to give up the flying that I loved or the midair recovery glory and business, I had to decide whether to re-enlist for just one stripe and be frozen in grade again or get out of the service. My final, brain-racking decision was to get out of the Air Force. So, our aircrew stayed intact after the Discoverer 14 recovery almost until I left Hawaii for home. I left Hickam AFB on the morning of December 7, 1960 (two years to the day after I first arrived there).

As I was processing out of the Air Force at Travis AFB I heard the news that the Discoverer 18 space capsule had just been successfully caught in midair by a Hickam AFB C-119J aircraft with the 6593d Test Squadron (Special). The plane was piloted by Capt Gene Jones. It was his second catch in less than a month and it was the third successful midair catch for the squadron and the Discoverer program. The plane consisted of an aircrew of ten, and upon their return to Hickam there was a jubilant celebration.

Memories flooded my mind and I knew then that it was going to be ever so hard to live with my decision to leave the Air Force. I was heartsick. Part of me wanted to go back to Hawaii and part of me wanted to go home. I was homesick for #037, our aircrew and the 6593d. I completed processing out of the Air Force and then I left California. I was on my way home to Pennsylvania.

Hill remembers the day he learned about the Corona program and the full story of the Discoverer space capsules the 6593d recovered. He gives a warm salute to his Air Force assignment, the aerial recovery mission, and his C-119 crew members.

One late night (in either late 1996 or 1997) I was glancing at our local newspaper, and also on occasion flipping through the television channels searching for something of interest. I stopped on the Discovery Channel. The program was about some type of intelligence activity, based on the content that I was hearing. Continuing to scan the newspaper and more or less listen to the television, it was only a short time until I heard the program narrator mention something about the 6593d Test Squadron (Special) out of Hickam AFB. With my ears perked up, I immediately tossed the newspaper aside, for now the television program was a great priority. I hurried, found a blank videotape and started recording this Discovery Channel program that had just started to bring back cherished and fond memories of the 6593d.

As I continued to watch, the program kept flipping from narrator to narrator and they were all talking about our old squadron, the Discoverer program and Corona. I thought, "What the devil is this Corona?" I carefully listened to every word and kept focused on the television. Then I saw a face that I recognized somewhere from the past, but I couldn't put a name to him or pinpoint a time of reference, only that I knew or had seen this individual in the past. Soon the program switched to someone else who I also thought that I had seen somewhere. I kept telling myself that I knew these guys from someplace, but could not put my mind to the specifics. Then the one person that was most familiar to me was mentioned as Everett "Andy" Anderson, a retired aircraft navigator who had been with the 6593d. At that point it dawned on me, that yes, I did know him from our squadron, even back to the days of our initial 6593d Test Squadron (Special) startup at Edwards AFB in 1958. Being a key navigator with the 6593d, Andy participated with the Operations Staff of the 6594th Recovery Control Group. He also coordinated operational matters with the various participants of the recovery efforts of the Discoverer/Corona space capsules. He was a special training project officer for recovery gear, directional finding equipment, and the crews on the aircraft.

The other person was later revealed as retired Col Frank Buzard who had been the Deputy Director and Operations Officer of the Discoverer/Corona program (1958-1963), where he had overseen the launch and operation of sixty-two Discoverer satellites. I probably remembered his face from the time that I was invited to Sunnyvale space operations or when I was invited to Andrews AFB to be a guest of Gen Schriever after our midair recovery of Discoverer 14. I won't go into what was said about the Corona program, other than it was the Top Secret code name, and the reason behind the filming of the Soviet Union and other countries during the Cold War days was to learn more about their missile sites and whatever could be determined by filming from our Discoverer space capsules as they orbited the earth.

The Corona program had been declassified by the White House on February 22, 1995, and the Discovery Channel program (sometime several years later) was about the Discoverer/Corona programs and some of the events that I had once personally been able to perform to help pave a path into history. The program was called *Secret Satellites*.

All those years went by since the 6593d Test Squadron (Special) and the Discoverer program were founded, to the special night that I watched this *Secret Satellites*. The speculations had now finally been answered that we and the news media had concerning the cameras onboard and the reconnaissance. I thought, "This is a fine way to find out, isn't it?" It was really good to at last find out what the truths of our missions and experiences had been about. I just wished that someone would have had the forethought to at least look me and other squadron members up and inform us of the true facts concerning our past services to the Discoverer/Corona program. But, what the heck? Down deep, we knew it anyway. We just didn't have all the accurate details.

I covered much of the duties, incidents and events of our 6593d Test Squadron (Special). The people were great to work with, the times and events were very interesting. I loved flying and surely enjoyed what we did, being a part of the Discoverer/Corona programs.

Daniel Hill with the Discoverer display at the National Air and Space Museum in 2002./Photo provided by Daniel Hill

I was very pleased when I finally knew that our Discoverer 14 space capsule had more photo coverage than all twenty-four of the U-2 reconnaissance aircraft flights combined. Our famous space capsule carried and returned with 1,650,000 square miles of film coverage of the Soviet Union. The Corona's vast geographic film coverage opened up the door and continued to bring great knowledge about the location and size of the Soviet Union's intercontinental ballistic missile sites and other military installations, thus being a turning point during the Cold War period.

Looking back into this period of my life, back 43 to 46 years ago, has brought up memories of numerous events and experiences that I am grateful to have taken a part in. Of course, had I not had the opportunity to do this interview (choosing to use e-mail) about my experiences with and about the Discoverer/Corona programs, I would have gone on with my normal everyday life, growing older, perhaps never to search the past adventures that I had with the 6593d. Having over 1,000 hours of flight time in our Old Lady, #037, and claiming 250 practice and live recovery flights with our aircrew, commanded by Capt Harold E. Mitchell, was a thrill and an extreme high point in my life. As I neared the end of my Air Force enlistment, the Discoverer 14 midair recovery certainly put the icing on my birthday cake. We had collected a lot of fame and made a little history along the line. Our 6593d was titled a "(Special) Squadron" from the start; it truly deserved that name.

> *"Our 6593d was titled a '(Special) Squadron' from the start; it truly deserved that name."*

Chapter 3

Capt Robert D. Counts

1st Lt Robert Counts holding his Aerospace Primus Club plaque and shaking hands with Lt Col Ahola in 1960./Photo credit: USAF

"We were the toast of the town in New York for a couple of days. It was a very big deal, and the Discoverer 14 recovery made international news."

Capt Counts (1935-) was interviewed on November 12, 2003. As a first lieutenant in 1958, Counts became one of the original members of the 6593d Test Squadron (Special). He was assigned to Capt Harold Mitchell's C-119J aircrew, and was the navigator who located Discoverer 14 during the first aerial recovery. He was assigned to the 6593d from 1958 to 1962. Counts was then assigned to Operating Location Number 1 (OL-1) at Edwards Air Force Base (AFB) from 1962 to 1968 as a C-130A and JC-130 navigator for aerial recovery flight tests. Here he developed the aerial recovery system for the Ashcan project and participated in a variety of aerial recovery flight tests.

Robert Counts was at Travis AFB when he received his orders to join the 6593d. He explains his path to the 6593d and how all of the squadron's navigators came from the Military Air Transport Service.

I was just a young lieutenant bachelor navigator flying for the Military Air Transport Service (MATS) out of Travis. My squadron had C-124s flying throughout the Pacific basin. One day, I went in after a flight and the adjutant called me into his office and said, "I want to talk to you for a moment." He closed the door behind me. I thought I was probably in trouble of some sort. He opened up the safe and said, "I have orders here for you. They're secret." He handed them to me, and I was to report down to Edwards AFB. Although the Air Force was supposed to give you a thirty-day minimum notice on a permanent change of duty, I had to be down there in six days. He didn't want to tell me earlier because it was a secret. Anyway, the 6593d formed nine crews down there at Edwards. I'd been a first lieutenant for about six months at that time.

It was a little strange the way the 6593d did this. First of all, we had nine complete crews and nine aircraft. The pilots (the aircraft commanders) were all captains, they were all married, and they all had been on previous aerial recovery projects, such as the Genetrix or Moby Dick projects. The copilots were all lieutenants, they were all single, and for the most part they had been on this previous project.

The navigators were all first lieutenants, they were also all single, and they were new to aerial recovery. They were all taken from MATS. The reason they did that is not real clear to me, except innuendo was that the pilots were not satisfied with the performance of the navigators in the previous recovery projects. I don't know any details about that at all, but they didn't sign on any of the previously experienced navigators. They probably took the navigators from MATS because they wanted people with worldwide experience. The MATS' bases in those days included: McGuire AFB, New Jersey, Charleston AFB, South Carolina, McChord AFB, Washington, and Travis AFB. They took two navigators from each of those, and an extra one from somewhere. I'm not sure where he came from. Between the two navigators from these four different bases, we'd been virtually every place in the world. I assume that's why they did that. I don't know exactly what their actual selection method was.

Counts describes the navigator training required for Discoverer recoveries. He explains how the homing equipment worked and the processes to recover a capsule. He also discusses the equipment issues of the C-119J.

We had to transition into the C-119, which was not much of a transition. It had the same equipment essentially that the C-124s and the other airplanes I'd flown had. The C-119 navigators operated the homing equipment that homed-in on the beacon of a descending capsule. Our homing beacon instructors were the civilian technicians who built this radio equipment, and they came there and taught us how to operate it. I want to say they were from ACR Electronics Incorporated, but I'm not certain of that, but they weren't All American. We had the ground support to show us how to maintain it.

1st Lt Robert Counts and his 6593d aircrew at Edwards AFB in 1958, left to right: (front) Capt Harold Mitchell, 1st Lt Robert Counts, Capt Charles Clawson, (back row) TSgt Louis Bannick, SSgt Arthur Hurst, SSgt Algaene Harmon, A1C Walter Johnson, A1C Billy Gurganious, and A2C Daniel Hill./Photo credit: USAF

Chapter 3 - Capt Robert D. Counts

A noncommissioned officer inspecting the homing antennas on a C-119J of the 6593d at Hickam AFB in 1960./Photo credit: USAF

The navigator always flew during training recovery missions, even though navigation duties weren't necessary. The training missions at Edwards AFB recovered practice capsules over the bomb range. There was no need to have a navigator, per say, but we always flew to either practice our homing skills with the direction finder, or as the safety officer, or for other things that had to do with navigating.

We had a lot of practice using the C-119 homing device. At Edwards AFB, we set a beacon on the ground and practiced flying back and forth locating its position. The beacon on the ground was mobile, and they hid it someplace out on the bomb range and we'd go find it. We were there at Edwards AFB for six months and then deployed to Hickam AFB.

There at Hickam AFB, we also did similar training missions. Then we had a series of several practice recoveries where a bomber would come flying out of somewhere in the States, and the whole reentry/recovery sequence would be simulated. We would be out in our recovery pattern and listening for the radio signal. Then the bomber (way up there out of sight to us) dropped a dummy training payload. It came down and it was a very realistic payload. It had everything on it that a regular Discoverer parachute and capsule had. The training payload had chaff that deployed and could be picked up by radar, not by us. The C-119 had no radar.

We had support from RC-121 Constellation aircraft out of Sacramento, California. Their main duty was coastal surveillance but they came out and helped us. If I recall, there were three RC-121s generally, and they had radars that picked up this chaff. They also had homing equipment that located the beacon and helped us vector in on the parachute.

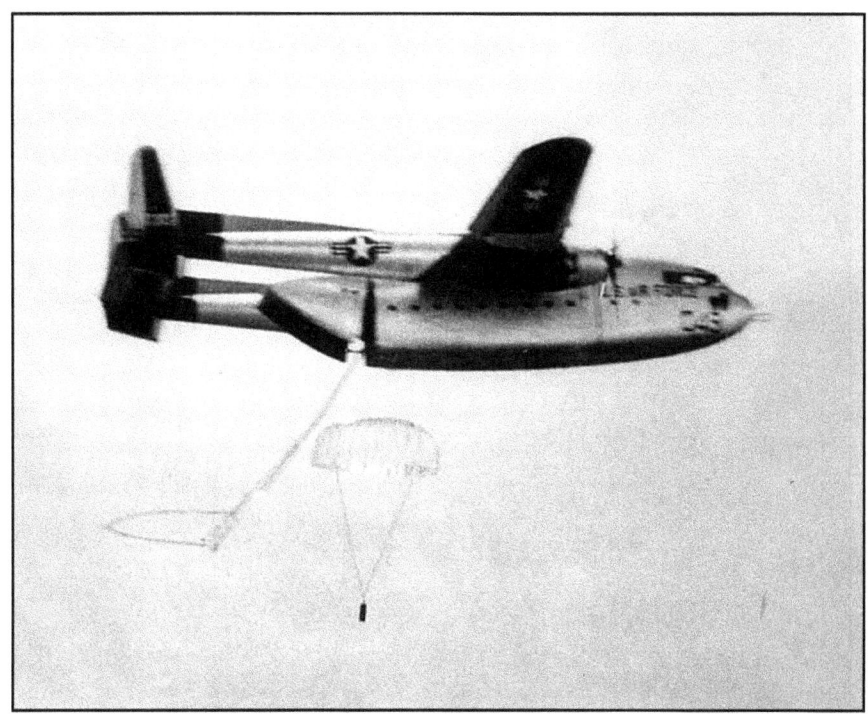

C-119J recovering a training parachute in 1959./Photo credit: USAF

The chaff deployed and a parachute deployed from the training payload. The radio came on and we homed-in on the training parachute, found it, recovered it, and brought it back. We did that several times as a very realistic practice. Of course, the first actual Discoverer capsule aerially recovered was Discoverer 14. So, we had Discoverer 1 through 13 to practice on, although a lot of Discoverer capsules didn't get that far. Often the Thor blew up on the pad or something.

The C-119 homing equipment was a little bit primitive. It was just two antennas mounted side-by-side on the nose of the airplane with reflectors. It looked like a modern television antenna. The signals from the antenna displayed on a screen. The homing equipment only told you if the capsule was left or right. It didn't tell you the distance to the parachute. If the capsule was to the right, then the antenna on the right got a stronger signal, and so forth. So you had the pilot turn the entire aircraft around, which turned your antennas, until you had a matched signal. Then you'd home-in on that signal, and you'd turn left or right a few degrees as you got closer, but eventually it led you straight to the parachute. It didn't give you range. It didn't give you elevation. It only gave azimuth, and only left or right.

The C-119 homing equipment had an ambiguity. If the capsule was straight behind you and on the right, you could think it was straight ahead of you and on the right. It had that problem. You could tell when you were straight on course, provided it was not straight behind, instead of straight ahead. You couldn't tell if it was behind or ahead of you. Except, if the parachute was behind you as you flew on, the capsule's signal would get weaker and that gave a clue that you were going the wrong way. Other than the strength of the signal, there was no way to tell distance. There was really nothing to compare the signal strengths to, so you weren't sure whether you were 10 or 100 miles away. That ambiguity problem caused us to miss Discoverer 13, which was the first satellite capsule to ever reenter the atmosphere, and was picked up by the Navy when it hit the ocean's surface. We failed to pick up Discoverer 13. If we didn't have that directional locating problem, we probably would have made an aerial recovery there.

As a navigator, Counts had the additional duty of being the flight safety officer. He explains the dangers of aerial recovery and what was required of a safety officer during a recovery.

The navigators had duties other than just navigating. One of those duties was being the flight safety officer. During a recovery, we didn't have any navigation duties to do. We were just flying in circles around the parachute. The navigator would go to the back of the airplane and stand out of the way and watch things. Hopefully, he would see any safety problem before it got too big.

I consider aerial recovery a hazardous operation. It's against the law to fly an aircraft within 500 feet of any obstacle. Yet that was our job, to fly within 4 or 5 feet of a parachute. There were a lot of potential problems, and many that actually came about: flying into a parachute and having it hang over the tip of the wing, or go into the prop, or whatever. The recovery itself was pretty hazardous.

There was a condition called an "inversion" where the recovery rig grabbed the parachute too low. The pilot had a very small window to hit on his approach to take these recovery hooks into the parachute. If he dragged the parachute along the belly of the aircraft and the hooks came into the parachute too low, instead of sliding into or initially grabbing the canopy, the hooks slid down and came to a stop at the payload. The parachute then just turned backwards and became a giant drag-chute. This inverted parachute had to be cut loose pretty quickly, because it was more drag than the airplane could pull. Usually, it was sort of self-canceling. An inverted parachute would just break the recovery line or something else, but the recovery would fail.

C-119J reeling in a training parachute and payload in 1959./Photo credit: USAF

An inverted parachute could be dangerous, because this was a nylon recovery line, and inverting the parachute stretched the line out to its fail point, which was about 30 percent of its length. The recovery line was 100 feet long, and if it stretched out to over 130 feet it would break. The stretched recovery line was like a rubber band that could fly back into the airplane with great force, and it could probably kill you. That happened a time or two on my aircraft. Luckily no one was hurt.

According to Counts, enlisted personnel did not have access to the same information officers did. He describes the program briefings he and other crew members received. Regardless, Counts and the other squadron personnel knew the content of the capsule was valuable.

The enlisted people had a disadvantage in that they weren't briefed on a lot of this stuff, at least not to the detail that the officers were. We knew pretty much what was going on. We knew when the capsule was launched. We'd go down to our group headquarters where they had radio and landline communications with all the tracking stations, the launch site, and all of that. We listened to things as they progressed to see if we were going to be a part of a live recovery mission, or if the Thor blew up and did not achieve orbit, or whatever. We were pretty well informed of the tactical part of it and what was going on.

They called the mission itself "Discoverer," and it was portrayed as a research and development program. But the officers of the 6593d knew what it was. We knew Discoverer was a reconnaissance satellite program. Many of the people in our unit had already been on the other missions, the Drag Net and the Moby Dick and all that. We knew it was aerial reconnaissance of the Soviet Union, and so forth, and we were getting the film back.

I don't really remember if we were briefed about the Discoverer's reconnaissance mission, or if that information got around informally. It was certainly general knowledge. I assume we were briefed on that.

C-119J of the 6593d at Hickam AFB in 1960./Photo credit: USAF

We didn't know a lot of the details. We didn't know how much film was in a capsule. We didn't know the technical things about the cameras or any part of that. But the officers of the 6593d knew the capsules had a large film cassette, and that it was very valuable, and it had to be recovered and protected at all costs.

Counts recalls some of the first Discoverer missions (1 through 13) and the problems they had. He also reminisces about the friendly competition between the nine squadron crews.

I guess that between Discoverer 1 and 13 we deployed for, say, five of them. I recall one, Discoverer 8, that everything seemed to work fine, except at the end when the parachute was to deploy. It just all happened too quickly. The capsule hit the water before anybody could begin to get to it. It was determined later on, as I recall, that the ceramic heat shield didn't come loose that protected it from the heat of reentry. It was supposed to come loose when the parachute deployed but it didn't. The heat shield added so much weight to the capsule that the descent was extremely rapid and it hit the water. So you could say that it was successful, all but that one little glitch.

There were a few Discoverer launches we deployed for that were thought to be alive and workable, but then something happened. Incidentally, Discoverer 2 was probably successful, but it came down in the wrong place. This is just a rumor. I have no way to know if this is true or not, but it was talked about, that Discoverer 2 actually deployed, not over the Central Pacific but over Spitzbergen, Norway. It came down but Spitzbergen was occupied by a lot of Russians. It could have been recovered by the Russians. It sounds like a long shot to me but that's possible.

The 6593d aircrews all wanted to be first to recover a Discoverer space capsule, for sure. But when you say "competitive" it was not competitive like, "I'm going to push you aside and get it." It was all well organized. The recovery pattern had all nine of the aircraft out in a north/south line.

The ephemeris of the Discoverer satellite, its north/south track across the earth, was very precisely known. There was nothing really that could make it deviate from that. The satellite was in a polar orbit. It came over the pole and headed south. A lot of variables could affect how far downrange the Discoverer capsule went after it started the reentry sequence. There was the angle of entry, and whether it spun up just right, and various things could affect its north/south reentry. We would be along the ephemeris, underneath the satellite.

We'd be along the ephemeris in a long line of nine airplanes. The primary recovery area was up at the top. That was probably where the Discoverer capsule was going to come in, but it could come in further down. So the aircraft were fairly, tightly spaced in the primary area. I don't remember the spacing. I guess it was 30 miles apart. They wanted to cover a lot of the secondary recovery area, but they couldn't cover everything, of course. In the primary area the spacing was such that no matter where the Discoverer capsule came in, at least two airplanes could get to it. In the secondary area the airplanes were much further apart, I'll guess at 90 miles apart. If it came down between the airplanes in the secondary recovery area, probably only one airplane could get to it. So who got to be in the primary area and who got to be in the secondary area was the big question, but the recovery area assignments were based on a rotational basis. If your airplane was in the primary recovery area this mission, then you'd be in the secondary recovery area the next time and so forth.

Harold MItchell was the aircraft commander and I was his navigator, and we were put in the primary area for Discoverer 13. We were Pelican 1 for the Discoverer 13 recovery mission. The airplanes' call signs were Pelicans 1 through 9. We were Pelican 1, the top recovery location assignment for Discoverer 13, but we failed to catch Discoverer 13. Since we were the primary aircraft of that mission, then we were going to be in the secondary area on the next recovery mission. We got rotated. Now we were Pelican 9, the very bottom recovery position for number 14. Well, as luck would have it, the mission was a week or so later. We caught Discoverer 14 as Pelican 9, way at the bottom, in the least likely place that the space capsule would reenter.

Due to a combination of system ambiguity and the poor radar reporting from an RC-121, the Discoverer 13 capsule landed in the ocean. Counts and the crew of #037 were disappointed. Counts recalls the day of the Discoverer 13 recovery mission.

The first space capsule to be recovered was Discoverer 13, and we were Pelican 1, the primary recovery position. When Discoverer 13 came in, as I mentioned before, we had these RC-121s that had all the radar gear on them. They picked up the capsule's signal that they thought was the chaff and started our C-119 towards it. After that I picked up the capsule's radio signal on my equipment. It seemed to be ahead of us just like the RC-121 crew was saying. So, we were borne off for it.

As we continued, I noticed the capsule's signal getting weaker and weaker. So I told Mitchell that I thought we were going in the wrong direction. So he inquired to mission control, "You sure this is it?" They said, "Yeah, we're sure." Well, now we had this ambiguity, and the RC-121 crew might have been sure about the capsule's location, but we weren't sure. Eventually, we broke off from the RC-121's recommended vector, turned around, and headed back the other way. Sure enough, as we progressed the capsule's signal got stronger, and I knew that we had been going the wrong way, but now we were headed the right way.

Maybe we saw the Discoverer 13 capsule hit the water. I'm not sure. If we didn't see it hit the water, we came across it just after it hit the water, but well after we could have recovered the capsule. We would have had to get to the descending capsule by about 2,000 feet. We didn't have any chance to get the recovery equipment out and have an effective recovery. Not recovering Discoverer 13 was a very big disappointment. Here was the first Discoverer capsule that could have been caught and we didn't get there in time.

We didn't get blamed for not recovering Discoverer 13. We always had a debriefing for all of our missions, and tried to figure out what we did, what went wrong, and what could be done better and so forth, but nobody criticized us for failure there, because we were doing what we were supposed to do. The RC-121 was the primary acquisition tool, so following their instructions was not the wrong thing to do. The aircraft commander of our airplane had full discretion to do what he thought needed to be done. If Mitchell had said right off, "No. We don't believe that heading is correct. Let's go the other way." We could have done that, not breaking any rules or anything. Given the information available at the time, I think Mitchell did the right thing.

It's speculative, but I'd have to say that our crew would have recovered Discoverer 13 if the RC-121 hadn't said anything to us, but the RC-121 crew might argue with that. The RC-121's tracking equipment was much better than ours in the C-119. They were equipped to look for Russian bombers and all that sort of thing. The airborne warning and control system early warning aircraft had the most sophisticated equipment available, and the RC-121 had a whole airplane full of it. It took a bunch of technicians to operate their equipment. Our airplane was full of recovery equipment. Our direction finding capability was primitive and small, and it was operated by the navigator who had other things to do, and our navigators were not nearly as proficient or experienced as these guys in the RC-121 were. So, that's why the RC-121s were there. The RC-121s had a much better tracking capability than we did, as far as capsule acquisition was concerned.

A Navy helicopter and frogman, Boatswain's Mate Third Class Robert Carroll, recovering the Discoverer 13 space capsule on August 11, 1960./Photo credit: USAF

The RC-121s continued to be with us for the next few missions. Eventually, they didn't support us anymore because things became more reliable and we didn't really need the RC-121s. I don't know how many recovery missions they participated in, but I'm sure that they, at some level, had to figure out what went wrong during the Discoverer 13 mission and why the RC-121 crew sent us the wrong way. There are all sorts of thunderstorms and other activities going on that could easily have been mistaken for, perhaps, the chaff. Just as individuals, maybe we were a little more leery of following the RC-121 vectors after Discoverer 13.

For the Discoverer 14 recovery, Counts and the crew were in the lowest priority recovery position, or Pelican 9. After the disappointment of missing Discoverer 13, the crew of #037 were shocked and thrilled that they were the ones who recovered the Discoverer 14 capsule. Counts describes their surprise and enthusiasm of the day.

> *"Now it felt like we were being punished and low and behold Discoverer 14 came down pretty much right to us."*

We went down in recovery assignment priority to Pelican 9 with long faces. We were down there patrolling south of Hawaii, tail-end Charlie, the Pelican 9 recovery position, milling around feeling like we didn't really do our job on Discoverer 13. Now it felt like we were being punished and low and behold Discoverer 14 came down pretty much right to us. The recovery was fairly routine because Discoverer 14 reentered

The Discoverer 14 space capsule recovery on August 19, 1960./Photo credit: USAF

and I got the capsule's signal. We turned the aircraft and headed towards the signal. I immediately saw the signal increase in amplitude, so I told Mitchell that we were headed for it. He questioned me with justifiable skepticism and said, "Are you sure about that?" I said, "Yes. I'm absolutely certain." So, on we went at full speed. We got a visual on the parachute and capsule. The details of that, the altitudes and all, Mitchell would remember that better. We saw it in plenty of time, got our recovery equipment set, and came around and made a pickup.

I have an artist's rendition of the recovery. It has our C-119 there and it's very dramatic with clouds all around, which was how it was. A lot of thunderstorms happened on most all of these missions down in that area, well south of Hawaii. The picture shows a destroyer on the surface and another C-119 hovering in the background. Neither was accurate. There was no other airplane that was able to get to Discoverer 14 before we recovered it, so there was no backup. The destroyers were all north of us, so we were there all by ourselves.

Discoverer 14 was a routine recovery. We pulled it in. The chief pole handler was a man named SSgt Al Harmon. They were pulling everything onboard and, of course, the lines came on the hooks. The hooks had to be guided through so they didn't grab anything as they were coming onboard. Then there was the parachute coming onboard. It was being thrown around by the wind and everything, going along at about 125 knots with the back door open. The pole handlers were pulling all this stuff in. Then the actual capsule came up to the back of the airplane. The pole handlers had to reach out and get the capsule, help it over the lip, and heave it onboard.

> "When Harmon touched the capsule, he jerked his hand back because it was hot. Then he touched the capsule again and it wasn't really hot, but it was quite warm."

When Harmon touched the capsule, he jerked his hand back because it was hot. Then he touched the capsule again and it wasn't really hot, but it was quite warm. It surprised him that the capsule was warm, and that is why he jerked his hand away. Then we grabbed the capsule and pulled it in. Harmon was the first person on earth to feel the heat of reentry. I know that might be a stretch, but anyway that always stuck in my mind. The first satellite capsule of any sort, or manmade object, to reenter from space was Discoverer 13, but it hit the water and the Navy recovered it. When the Navy got there the capsule was cold.

Recovering Discoverer 14 was a big surprise. I don't know exactly how to describe it. It was like a gift from the gods to let us have a chance to vindicate ourselves. We were very surprised that it happened, surprised and pleased.

After the successful recovery of Discoverer 14, Counts, along with Capt Mitchell and TSgt Bannick were ordered on a public relations tour. They stopped in Los Angles, New York, Washington, D.C., San Francisco, and then back to Honolulu. They were treated like celebrities as they toured the country. Counts gives details of the assignment.

After we landed at Hickam AFB, three of us were told to go home, pack, and come right back. That included Mitchell (the aircraft commander), myself (the navigator), and TSgt Louis Bannick (the winch operator). The three of us immediately went on a public relations tour. So we went back, took a shower, and showed up.

An hour later we were on a civilian airliner headed for Los Angeles. We flew first class, and were served champagne and all of that, all courtesy of the airline. I think it was United Airlines. We landed there at Los Angeles International Airport and we were interviewed in the civilian terminal. The people from the Air Force Ballistic Missile Division were there. The press contacted my mother so she was there to greet

Gen Emmett O'Donnell awarding the Air Medal to 1st Lt Robert Counts at Hickam AFB for his participation in the Discoverer 14 recovery. The photographed members of the Pelican 9 crew (left to right): Capt Harold Mitchell (aircraft commander), Capt Richmond Apaka (copilot), 1st Lt Robert Counts (navigator), SSgt Arthur Hurst (flight engineer), TSgt Louis Bannick (winch operator), SSgt Algaene Harmon (loadmaster), SSgt Wendell King (photographer), and A1C George Donahou (loadmaster). Not seen in this photo are A2C Lester Beale (loadmaster) and A2C Daniel Hill (loadmaster)./Photo credit: USAF

Lt Gen Bernard Schriever presenting the Aerospace Primus Club plague to 1st Lt Robert Counts in 1960.
/Photo credit: USAF

The Discoverer 14 capsule in 1960 being displayed at the Space Technology Laboratories' Research and Development Center headquarters building (Building 105 of Los Angeles AFB in 2006).
/Photo provided by Northrop Grumman.

Chapter 3 - Capt Robert D. Counts

us, along with Gen Bernard Schriever who was the commanding general of the Air Force Research and Development Command that we were under.

The Air Force made quite an event out of it. The public was aware of the Discoverer program in a general way. There were articles in the local newspaper about us and what we were doing once in awhile. For public relations reasons, the Air Force wanted to make the Discoverer 14 recovery known.

From Los Angeles, we flew to New York. We were the toast of the town in New York for a couple of days. It was a very big deal, and the Discoverer 14 recovery made international news. In New York, we were on all kinds of television talk shows and the national news. This was 1960, so I don't remember who the hosts were, but we were on NBC, CBS, and all that sort of thing. They were the big shows. I do remember a children's show in the morning. I would say it was *Captain Kangaroo*, but I'm not absolutely certain. The other guys didn't really want to go, so they sent me solo on that one. I enjoyed it very much. That was the essence of making a big deal out of the Discoverer 14 recovery.

From there, we went down to Washington, D.C., and we were there for a couple days. In Washington we were supposed to meet with John Kennedy, but it didn't work out. I don't remember the details of what happened, but Kennedy was not the president at that time. It was August of 1960, so Eisenhower was the president and Kennedy was a presidential candidate. We didn't meet either one of them. Then we went out to San Francisco. The Lockheed people there at Sunnyvale, who built Discoverer, took us out on the town in San Francisco.

After we went back to Honolulu and got back to our regular duties, I was asked to give talks on the Discoverer program to small groups: the chamber of commerce, the Lion's Club, or whoever wanted to have a speaker. I was the guest speaker at a lot of events. They gave me Discoverer 14 as a prop. I had the actual capsule and its parachute in a big container, and I carried this around in the trunk of my car from place to place. It was 1960 and the capsule was insured for a million dollars. I thought that was very generous of the Air Force to let me use the actual capsule and parachute to help me in my presentations. Everybody wanted to touch the capsule. My little speaking tour was only in Hawaii.

The capsule was gleaming gold and about the size and shape of a kettledrum. As I recall, there was some speculation in the press that the government was crazy for sending the National Treasury into orbit by having this capsule made of gold. One of the engineers pointed out that this was just a film of gold on the capsule as a radiation and heat reflector, and the actual gold on it was probably worth only about seventeen cents because it was very, very thin. But the gold looked good.

I gave the capsule back after I finished my touring. I don't know its eventual location. There are three of them in the National Air and Space Museum in Washington, D.C., and Discoverer 13 is there, but I'm not sure where Discoverer 14 is.[1]

Soon after the Discoverer 14 recovery, security tightened on the program. The Air Force no longer publicly announced launches or recoveries.

It wasn't too long after the Discoverer 14 recovery that the Air Force clamped a big lid of secrecy on the Discoverer program. It got very much tighter, which was a little strange since the information was already widely known. I suppose the Air Force didn't want the Discoverer program to be that much in the public mind, and didn't want people to know anything, like our schedule or how many recoveries we were doing. The Discoverer program was put on high security and became a national secret, even though, if anybody was reading the papers earlier they knew all about it.

---

[1] The Discoverer 14 capsule is on display in the National Museum of the United States Air Force at Wright-Patterson AFB, Ohio.

 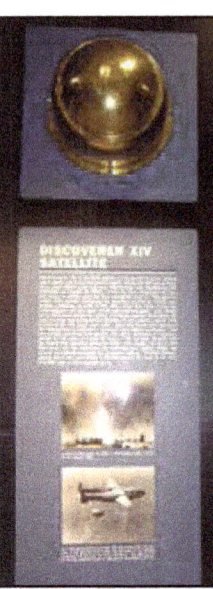

The C-119J that recovered Discoverer 14 and its space capsule on display at the National Museum of the USAF in 2006. /Photos taken by Robert Mulcahy

I was involved in many other Discoverer recoveries. The Discoverer 14 recovery was in 1960, and I spent another two years doing general recoveries. I don't recall how many space capsules I was involved in recovering. I suppose three or four, not very many. It became kind of routine. These actual recoveries were flying in the C-119 and later in the C-130. We transitioned over to the C-130s while I was there at Hickam. Actually, I probably flew a lot more recovery missions in the C-130 than I did in the C-119.

When the Discoverer program began, the C-119 was the best recovery aircraft available. When the JC-130 replaced the C-119J, the squadron was quite pleased. Counts describes the challenges with the C-119J and the improvements of the JC-130.

The C-119 was the best aircraft at the time. It had been used in those earlier recovery projects, the Genetrix and Moby Dick projects. The recovery equipment was developed for those earlier projects, so we used the C-119 pretty much as they had left it. These aircraft were given to the Air Force Reserve after the Moby Dick project and all the recovery equipment was taken out of them, of course. The C-119s had been acquired for the 6593d from the Air Force Reserve. All American Engineering developed the recovery equipment, put it back in our C-119s, and then we trained with it and so forth.

The C-119 was a good airplane for its recovery purpose at the time. It was relatively slow in maneuverability and speed, so it could get down to a lower airspeed to make a recovery. It had high wings and high engines so you wouldn't have too many parachutes go into props during recoveries.

The most important C-119 feature, of course, was its air-operable cargo door. The Air Force had two models, the clamshell door and the beavertail door. We had the beavertail; you could open the door in flight. If a C-119 was going to make an airdrop with the clamshell doors, the aircrew had to leave the doors behind when they took off. That wouldn't have worked very well for us, because it would have decreased the aircraft's range and all of that.

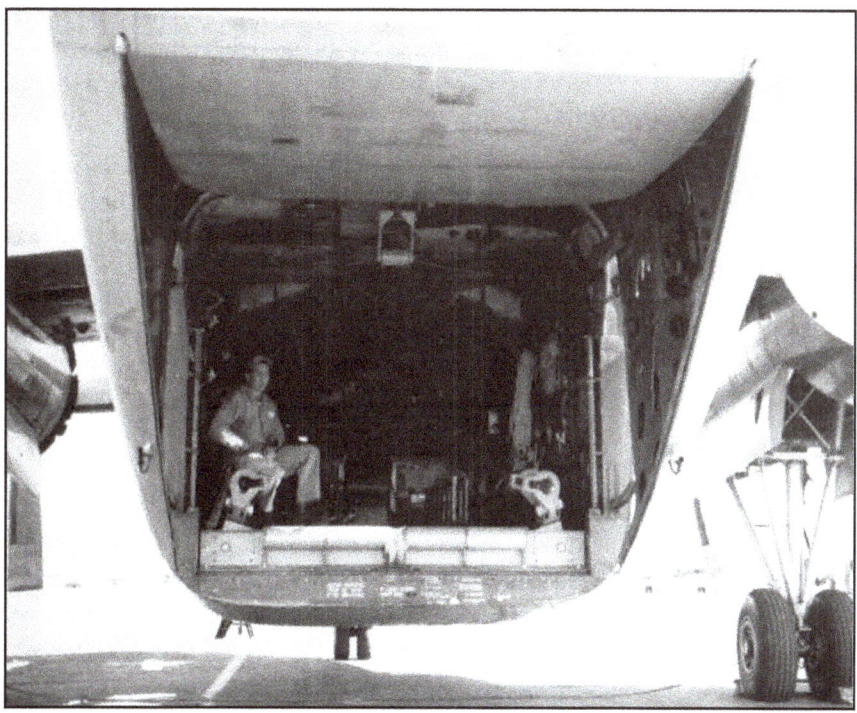

A beavertail door on a C-119 that could be opened upward while in flight to recover parachutes. /Photo credit: USAF

JC-130 recovering a test package and parachute in 1962. The package is probably an expended jet-assisted takeoff (JATO) bottle filled with shot. JATO bottles were often used during flight test recoveries at Edwards AFB./Photo credit: USAF

The C-119 had some deficiencies. Its range was relatively marginal. Just to get from Travis to Hickam Field for our initial deployment in 1958, we had to wait around for several days, because we needed at least a plus 14-knot wind factor, as I recall. We had to average 14 knots of tailwind just to make it. Another problem was the C-119 had two engines and it wouldn't fly on one, so it was not the right airplane for long-range over-water operations.

Another deficiency was the C-119 would not ditch. There had never been a successful ditching of one. The C-119 always broke up. We had special air-droppable survival equipment. We had a pallet in the back loaded down with a life raft and a mixture of things that we'd need if we had to ditch. The rule was if you were at high altitude, lost an engine, and couldn't maintain altitude, you slowly sank. Then you would start throwing everything out the back trying to make it back to an airport. If you sank to 1,000 feet and couldn't maintain altitude, then you'd bail out, taking the pallet with you. You'd never ride the airplane below 1,000 feet, no matter what. We never had that problem. As a matter of fact, in all the time we had the C-119s, only once did somebody lose an engine, and that was right at take-off, close to base. They made it around and landed all right. The fact that the C-119 wouldn't maintain altitude on only one engine, and it would not successfully ditch, were two things that made it not very desirable for long-range, over-water operations.

We were very glad to come out of that unscathed. I have to compliment the 6593d maintenance personnel. They did an almost supernatural job to keep our C-119s flying all those years, all those thousands of flying hours, and have only a single engine problem. But you can only do that for so long.

The C-130 navigation equipment had all sorts of improvements over the C-119. The C-119 had an astrodome plastic bubble on the top of the airplane and a handheld sextant. You'd get up on a ladder and poke your head up in there, look at the stars, and take a sight. That was the same equipment they had in World War II. The C-130s had a periscope sextant that just poked a little tube through the skin of the airplane. It was much more accurate and easily manipulated, so that was a big improvement. We had radar in the C-130; the C-119 didn't have radar. We had upgraded Loran. We had Loran A in the C-119 and Loran B in the C-130, which was a big improvement. That was way before satellite navigation and GPS and all those wonderful things.

In 1962, Counts transferred to Edwards AFB and was assigned to the OL-1 of the 6594th. He was the navigator for their one crew. They were assigned to conduct research and development of aerial recovery missions. They also recovered telemetry data capsules from a ship and delivered them back to California. Counts describes these missions as well as his other job as a test engineer.

I left Hawaii in 1962. From Hickam AFB, I went to OL-1 of the 6594th Aerospace Test Wing at Edwards AFB to be in the research and development of the same aerial recovery program. I was the navigator for OL-1. They only had one flight crew.

We were doing the research and development for new equipment to make the Corona recoveries more reliable. In the earlier days, we at Edwards AFB went back to Hawaii with a C-130 for most of the live Discoverer missions, because we had an aircraft with all the recovery gear on it. We always flew a C-130. The C-119 couldn't make that flight as a routine thing.

The first test bed C-130 aircraft for the transition from the C-119 was a C-130A model, tail #53-3131. It had the same equipment as the C-119 and was tested and qualified at Edwards AFB to work out all the problems. That C-130A flew over and augmented the nine C-119s that were in Hawaii, so then there were

Chapter 3 - Capt Robert D. Counts

The recovery of Discoverer 26 on July 9, 1961./Photo credit: USAF

ten airplanes during recovery missions. Later on, even after they transitioned into the C-130s, we'd still take our airplane over to Hawaii and augment the 6593d from time to time, just to give them a bigger force.

When there was an operational mission, we flew over to Hickam a day early and were there to augment their recovery force. Later on, the 6593d had more airplanes and enough resources so we didn't have to go back to Hawaii. Even still, we'd fly back and get a recovered Discoverer capsule at Hickam, bring it back to the mainland, and turn it over to the people there at Sunnyvale. We did that quite a bit.

Another responsibility we had at Edwards AFB was to recover data capsules from the downrange ship. Through all this, there was a ship south of Vandenberg recording telemetry from the Discoverer capsules and the Agena spacecraft. The ship was right under the point of the Agena's assertion into orbit, about 1,500 miles downrange from Vandenberg, south of the tip of Baja, California. In a routine Discoverer launch there wasn't any need for us to make a data recovery from the telemetry ship, but if there was a failure of some sort, they had telemetry data there that they wanted to get to Sunnyvale as soon as possible. It would take the ship days to get back, so we flew from Edwards AFB.

An instrumentation ship (USNS *PVT Joe E. Mann*) in 1960 that supported the Discoverer program. The ship was positioned between the tracking stations at Kaena Point, Hawaii and Kodiak, Alaska and received telemetry from the Agena and the reentry vehicle during a Discoverer mission./Photo credit: USAF

We'd land at North Island Naval Air Station there in San Diego, refuel (at least in the C-119s), then we'd fly down to the ship, snag the data capsule off the ship, and bring it back. In those days, we'd have to refuel there at North Island and then take the capsule up to Sunnyvale. Later, the C-130s had the range. They'd just fly the entire route, more or less, out of Edwards AFB and pick the capsule up at sea and take it to Sunnyvale. We did a lot of research and development on those capsule pickups off the ship.

I remember one of those recoveries in particular. The ship was a little too far away for us. I thought we couldn't make it in a C-119. I was the navigator on that one, and I calculated that we were going to be about 20 miles short. I said, "We won't make it back." They said, "Well, go anyway." Because in all these calculations, tables, and performance charts, anytime the navigators rounded something off, they always calculated the range a little bit on the safe side. The navigator calculated a little extra fuel for the wife and kids and tried to always make it safe. They decided, "No. Just do the best you can. Get the airplane back as close as you can, but we need the telemetry capsule."

So we flew down, but the flight had to be just right with no unnecessary maneuvers. The ship was ahead on the horizon so we started our descent, deployed the gear, and talked with the ship. In those days, the ship trailed a line off its stern to a sea anchor, and then the capsule was attached to the line on the fantail. We'd come by and snag the line and the capsule, and make a big pull up. Normally, we'd make circles around the ship until we got the capsule in and secured, in case anything went wrong we'd lose it near the ship, and then we'd head back. On this one, we just made the descent, crossed the line, snagged the capsule on the first pass, made a u-turn, and headed for San Diego. We still had fuel over North Island, just like they thought.

The trailer off of the stern worked all right, but it had its difficulties. So they developed a balloon program. The ship's crew inflated a balloon, it went up in the air off the fantail, and we'd pick up the balloon with the airplane. It was much safer and more reliable than picking up this line just a few feet above the water.

My primary job at Edwards AFB was to be a test engineer. I conducted tests on various modifications to the recovery equipment, and tested new systems and requirements for aerial recovery. The Corona program was the biggest recovery effort, but there were a lot of smaller recovery projects in those years. Literally dozens of operations had things that they wanted snatched out of the air. They brought whatever it was that needed to be recovered to Edwards and a lot of what they brought had to be tested.

The Ashcan project was very important to Counts. He helped engineer the recovery system for Ashcan using his experience with Discoverer. It is clear Counts took pride in Ashcan. He discusses the project, the missions, and his hands-on role in the program.

The Ashcan project was probably the most unusual recovery project. The Air Force was doing high-altitude air sampling to see, from a scientific point of view, how particles exchange and eventually end up distributed around the world. Ashcan was a high-altitude air sampling device carried aloft by a balloon. They had the same balloons as the Moby Dick project and many of the same balloon launch and support people. When those projects ended, there was a lot of equipment that was just put away in a warehouse somewhere, and now, it'd been taken out.

Initially, the Air Force was doing the Ashcan project in Texas and they were taking air samples at mid-latitudes. They just let the payload come down and impact on the desert. Then they'd land the chase plane loaded with a jeep to pick it up and bring it back. Later on, after they had worked out their sampling problems, they needed to expand this to get samples at the Arctic Circle and the equator to get a full idea of what was going on in the upper atmosphere.

Anyway, the Air Force had done their Ashcan sampling and their work at 45 degrees without our help. But over the tundra in the Arctic Circle, they couldn't allow the payload to hit the ground because the sample would be contaminated. Over the equator, the same deal. Whether the payload landed in the ocean, or landed in the jungle, or wherever it landed, the sample would be contaminated. So the Ashcan project needed aerial recovery in those two locations.

The Air Force brought the Ashcan package to us at Edwards. They only had the air sampler. It was a long bar suspended by the balloon, and then out on the ends of this bar would be their payloads balancing each other on each side. This was a payload that would never stand the force of aerial recovery and would never trail in a stable way behind the airplane. They had no parachute system, so we had to develop that from scratch. We did that in-house. The parachute design was my own, more or less. That whole recovery package was my responsibility.

We had two problems. We had to repackage the Ashcan payload in a way that was compatible with aerial recovery, both in trail and with the forces involved. We also had to design a parachute for it that was compatible with aerial recovery. The trouble was, the Ashcan people wanted to deploy it in ten days. This project would take, I would say, a minimum of ten months if it went through the regular process. We decided to just jump on it and develop a recoverable Ashcan system right there, right then, at Edwards AFB.

The solution was to take our already proven parachute used in the Corona program and double it up. The Ashcan payload weighed about 300 pounds, which was much too heavy for the Corona parachute. We put two parachutes in tandem, one attached to the payload and the other parachute was above it with a line coming down through the apex of the lower chute to the payload. Both chutes were identical in size and construction. We came in with a C-130 and picked up the upper chute and recovered the whole thing—the payload and both parachutes.

The parachute system was packed on the Ashcan payload sort of like they deployed it. The lower chute was already in the deployed position, but the upper chute and the line connecting the two were packed into a parachute bag. When the balloon broke on command and dropped the payload, the bag was attached to the balloon. So, the bag stripped off the upper parachute and line as the balloon was left behind. The payload fell away, stripped out all the line, and then stripped out the upper parachute. The two parachutes deployed and it came down. We had to use off-the-shelf materials to do that. We had to repackage the payload, but we were able to design a little gondola out of aluminum tubing and deployed it in the ten days given.

We did the Ashcan tests there at Edwards, and then we did the operational missions as well. We'd deploy in the summer up to Eielson AFB, Alaska, for each of their missions. Then we'd deploy in the winter down to Howard AFB, Panama, in the Canal Zone, for their missions and did that for a period of time. I'll guess I was involved in the Ashcan missions for about two years, so that was about 1965 when that came to us. I don't remember the dates.

They launched an Ashcan balloon from either Eielson AFB or from Howard Field. It would be launched early in the morning. This was a giant balloon. The biggest balloon ever built to this day, as far as I know. It was about a million cubic feet. It covered a football field if you spread it out. It was a very big balloon.

The Ashcan balloon had to be launched in zero wind conditions. It could tolerate maybe 1 or 2 knots of wind at launch time, at the most, because they had to fill the balloon and it took awhile. They had a couple of semi-trucks with helium tanks on them. It took a couple of trucks to fill this balloon with compressed helium. As I recall, it took about an hour to fill. At first, the balloon was just a little ball of gas. They'd have the balloon stretched out on a long line and the payload itself was on a truck.

They started to inflate the balloon and this little ball of gas started to raise the top of the balloon. It slowly rose, and rose, and rose, until the balloon finally came up off the ground, but they still had to get a certain volume of gas into it. The balloon was there in this vertical position for a little while as they put more gas in it. They couldn't have any wind because it would blow the balloon over. The balloon was just made out of thin polyethylene, and was not all that strong.

They would finally get the Ashcan balloon in the up position. Then the launch truck with the payload had to get directly under the balloon. It never came fully straight up on its own because of what little tiny breeze was blowing. There was a driver in the truck and a guy in the back, and they both wore crash helmets, headsets, and so forth, so they could communicate. The Ashcan payload was held down in the truck. It was locked in the truck by a big arm that went across the very bottom of the balloon. The guy in the back told the guy in the front, "Steer the truck a little right, a little left, no, move faster." The truck driver lined up so he was directly under the balloon. He could not see. He had to have the balloon vertical. They drove around over the airfield trying to get the inflated balloon perfectly upright.

Finally, when they got the balloon just vertical, the guy in the back released the lever and the balloon started up, lifted the payload off the truck, and away it went. If the balloon was not vertical on release, it

Mark 8 parachute with an unknown payload. The JC-130 would strike the small white parachute with its recovery gear when recovering these parachutes.
/Photo credit: USAF

would swing like a pendulum and hit the ground. They usually started inflating the balloon at night, just before dawn. A Federal Aviation Administration regulation prohibits a balloon from flying at night. They had the balloon's preparation all timed with the target of launching the balloon right at sunrise. That's usually how it worked.

We flew our C-130 around in a circle watching the balloon's inflation. Then we accompanied the balloon as it made its ascent. Finally, the balloon left us behind, and we watched as it went on up. The Ashcan payload had a radio beacon on it as well and we tracked it. Then the balloon eventually reached altitude.

The Ashcan balloon went up 100,000 to 130,000 feet, and stayed up there for ten hours or so, with its blowers pumping air across filter paper. Even at altitude (135,000 feet is about as high as it would go) you could still see it from the ground. The balloon was so big that if you knew where to look, you could see it. It was just a silver spot. That was the reason for many UFO sightings in the 1950's during the Moby Dick development.

We watched the balloon visually and went with it wherever it went. Sometimes we recovered the parachute and payload over Panama. Sometimes we'd get it over Colombia. Sometimes we'd get it over

JC-130 and recovery crew in 1962./Photo credit: USAF

Costa Rica, but wherever the balloon went, we were there. We had diplomatic clearance to do all of this. It was all arranged ahead of time, so we could fly anywhere we needed to go. We also had a liaison officer from the Panamanian Air Force onboard to act as an interpreter in case of any problems.

The payload was up sampling for ten hours, so we had to fly around in circles for those ten hours until the sample had enough volume of air going across the paper. They needed to get the filter paper back totally intact and laboratory clean, so the Ashcan's payload doors closed and the filter paper was put away in the Ashcan equipment. Then we started the payload's descent.

We had equipment in the C-130 that sent a signal to release the Ashcan payload from the balloon whenever it was appropriate. The way the payload release worked, there was a tape attached to the payload that went up the side of the balloon; it was similar to the strip on the cellophane that you rip to open a new pack of cigarettes. When the payload fell away, this tape stripped one gore of the balloon. It just slashed the balloon from bottom to top as the payload fell away, breaking the balloon into little bits. It was so cold at that altitude that the balloon's polyethylene was quite brittle and broke into little bits (was the theory). Anyway, nobody ever noticed the pieces of balloon. Then the whole Ashcan payload separated from the balloon and descended by parachute. The parachute came down and the upper one deployed. We were tracking it in the C-130 as the parachutes and payload came down. Then we picked it up when the parachute was low enough.

There was one Ashcan operation that I recall when we were over Mount David, an extinct volcano in northern Panama or southern Costa Rica. The parachute came down pretty much over that mountain and we recovered it at 10,000 feet or so. The terrain there was at about 7,000 feet, and there was total cloud cover below. We had to recover it at a higher altitude because of this cloud cover over the jungle. The volcano was all covered with tropical jungle, even at that altitude. The clouds were so thick that you couldn't see the jungle and we had to be above the clouds, of course. We picked the parachute up and we were going around in circles as we were reeling the payload in. All of a sudden, I looked over and here's the darn balloon. It hadn't broken into bits like it was supposed to. During its descent, the lower half of the balloon had pushed up into the upper half and here it was coming down like a giant silver jellyfish 100 yards in diameter. We watched the balloon as it descended into the clouds and disappeared. I've often wondered, if we went back up there today, whether there'd be guys running around with pants made out of polyethylene and having some sort of a new religion.

We at Edwards AFB deployed for all the operational Ashcan missions. Then eventually, after I left, the Ashcan recovery mission was turned over to the 6593d in Hawaii, and they did the operational missions. The Ashcan project didn't last too much longer after that.

> *"My involvement in the Ashcan project was the high point of my Air Force career. As far as I know, Ashcan was very successful."*

I was involved in all the first Ashcan recoveries. I was always sent because I was the engineer who designed the Ashcan system, and if something went wrong I'd be there to deal with it. I was also a navigator on virtually all those recovery missions. I guess that I was on five or six Ashcan recoveries in Alaska and five or six in Panama. I don't know the total number of missions. I left in March of 1968. My involvement in the Ashcan project was the high point of my Air Force career. As far as I know, Ashcan was very successful.

For Counts, Corona was only the beginning of his time in aerial recoveries. He worked with several other recovery programs. He engineered the Ashcan recovery system, worked with National Aeronautics and Space Administration (NASA), assisted the X-15 program, and supported the Apollo missions. He even had a hand in "UFOs." Counts recalls his time with each program. He holds his time with the 6593d and Corona high. However, it was the Ashcan project that he claims as the pinnacle of his Air Force career.

The X-15 program at Edwards AFB had equipment that shucked off of the X-15 while in flight. They wanted to improve the performance of the X-15, so they put some external belly tanks on it, much like the space shuttle has on it at launch. For aerodynamic reasons, they had to put a ventral fin, a lower fin, on the tail of the X-15 for control. The X-15 expended the fuel in the external tanks first and then they dropped away, just like the shuttle. The fin would shuck off because it was in the way for landing. All this was fairly valuable stuff so they wanted it recovered. The external tanks came off first. They came down by parachute and impacted on a large dry lake near the launch site in northern Nevada. The ventral fin ejected later and was recovered by us before it hit the ground. They retrieved the tanks. We were involved in the development of those systems, as well as the operational recovery of the fin.

We were involved in a lot of NASA/Air Force launches at Cape Canaveral/Cape Kennedy, Florida. Various cassettes and pieces of equipment were shucked off their manned and unmanned programs and so forth. They came down by parachute. We'd pick them up and bring them back. These programs were small, one-time efforts, the details of which I don't recall.

We were involved in the manned Apollo space capsule recovery. Initially, there were three different recovery methods that were considered. One was to just have a big parachute and let the space capsule flop down in the sea. That was the recovery method that was finally adopted by NASA because it was deemed to be more reliable. We were involved in the other two recovery ideas. One was to recover the capsule in the air and bring it in, just like we did with the other programs. I think in the end, the danger of us failing just added to all the other dangers. Although, in my mind, aerial recovery was the way NASA should have gone, because they would have gotten the Apollo capsule back laboratory clean, not dipped in the sea.

Our recovery effort for the Apollo program had the same general technique as the Moby Dick project for those heavy payloads. It was a large main chute and then a small drogue chute coming up from that, a tandem chute again. When we tested for recovering Apollo space capsules, we used a great big parachute and a tiny one, and the tiny one was our recovery target. If we failed to recover the target parachute, ripped through it or whatever, the astronauts would still have the main chute and could land in the ocean anyway. But there was some possibility that we could snag the parachute, get it half in, and then have something happen to add a certain amount of additional risk, but not much.

The third recovery program was for the Apollo capsule to deploy a Rogallo wing like a hang glider, instead of a parachute. The Rogallo wing unfolded and then could be controlled from the capsule. The astronauts would actually fly the capsule with the wing above the capsule. The capsule shifted its weight around through the suspension lines from the hang glider. Then the pilot astronaut, by reeling in with electric motors on the cables, could shift the center of gravity around, maneuver the capsule, fly it back, and land it on the aircraft carrier. So, all this was worked out and we were doing these tests at Edwards AFB.

We dropped the capsule from a C-130. The Rogallo wing deployed and glided down. A NASA test pilot in our aircraft had a radio control, a little joystick there in our cockpit, and flew the Rogallo wing-supported capsule around. The first capsule flight worked OK because we did it early in the morning. Like everybody else, we wanted quiet thermal conditions and still air. The wing-supported capsule flew just fine.

We did the next flight in the afternoon. That didn't work nearly as well because the maneuvering Rogallo wing ran into a thermal and started to bounce a little, get a little up and down motion to it. Then these long suspension cables going down to the capsule alternately got slack and taut. So they'd go slack and the wing went into some type of a weird turn. The pilot tightened up the cables and tried to correct the capsule's flight, but he was just cranking on slack cables. It became taut again and he overcorrected it way too much or whatever, anyway the capsule crashed. It went out of control and crashed. It was analyzed and the thermals were the trouble. Over the ocean, maybe you don't have a lot of thermals, but you have thunderstorm activity, so that whole flying capsule program was scratched after a great deal of effort.

> *"Civilians knew I was in the Air Force at Edwards AFB, and from time to time they said, 'Tell me about UFOs.' I would say, 'I know all about UFOs. I've even flown in them.' Here's the story behind that. I wouldn't tell them 'the' story. I just left them wondering."*

Chapter 3 - Capt Robert D. Counts

Robert Counts (right) and Daniel Hill (one of Counts' C-119 aircrew members in the 6593d) at the National Air and Space Museum in 2002./Photo credit: Daniel Hill

Civilians knew I was in the Air Force at Edwards AFB, and from time to time they said, "Tell me about UFOs." I would say, "I know all about UFOs. I've even flown in them." Here's the story behind that. I wouldn't tell them the story. I just left them wondering.

There was a requirement to develop a nighttime parachute recovery capability. All sorts of things were thought of and tried at Edwards AFB. Things like having a big spotlight on an airplane so you could fly around and shine it on the parachute. We used a little two-engine, two-tailed Navy aircraft for that, a sub hunter. I don't remember the designation of it. It had a giant spotlight on the wing to visually search for things on the surface at night by the Navy. So, we borrowed one of these. Our pilots flew with Navy pilots, a copilot/pilot sort of situation, to use their spotlight but our expertise in recovery. It didn't work. The trouble was, we could see the parachute but there was no background. It was just a parachute in a black sky. Is the parachute above you? Is it below you? Without any reference it was very difficult to tell. It was decided that it was just too dangerous to try to get that close to a standard parachute at night.

The next effort I recall (there were probably others) used a strobe light on the capsule itself. It shined up into the parachute canopy. It had to be a strobe because it had to ride in orbit, and pounds in orbit are very expensive. A strobe would last and do a lot more for its size than a regular spotlight.

To give it some continuity, the parachute itself had an aquadag coating. If you're not familiar with aquadag, it's a florescent coating. It's the same thing that's on the tube of your television set. When you shine a light on it and take the light away, it fluoresces for a moment. This strobe went *flash, flash, flash*, but between flashes the canopy was still glowing, so the illumination was sort of constant. It was flashing but there was still a residual constant target. The strobe had the same problem as the spotlight. The pilot wouldn't have any altitude references.

Anyway, so we were going to test this nighttime parachute recovery over the bomb range there at Edwards to see if this was going to work. We threw the parachute and strobe out of a C-130. The parachute opened up. It was going *flash, flash, flash* and glowing and coming down like a parachute. We've got our strobe going on our C-130, and our red running lights and our green navigation lights, and all of that. We were flying away from the glowing parachute then coming back. We took a pass over the parachute just to practice, and then another pass to see how it's working, and then another pass over it to see how it works and figured, "OK, the system is alright." So we came in and picked up the parachute. We flew around, reeling the parachute in. Our strobes on the aircraft were going *flash, flash*. The parachute strobe was going *flash, flash*. We brought the parachute inside the C-130, closed everything up and we went home. I'll say that was in 1968, somewhere in there.

The next day the San Bernardino morning newspaper had a big headline about UFOs that were seen over the desert. Some guy was driving on Highway 395 that goes by Edwards AFB, and it was midnight, or something, when we were doing our test. He was looking over at the bomb range and saw this recovery test going on, and he described it almost to a T. His description was very accurate, but his conclusion was totally off, because to him, our airplane and the parachute were UFOs. One "UFO" was going here and the other was going there, and all of a sudden they merged, and one seemed to go inside the other, and then they both disappeared. Like many UFO stories, the witnesses think it's a UFO.

We worked on that for many nights. The parachute recovery program was a military secret. The Air Force was not going to call the newspapers and tell them, "No. You got all this wrong. That was really a classified military program, but don't tell the Russians this." So we just let the UFO article go and let the newspapers run with it. We would just forget it. So, that was the time I flew around in a "UFO." Of course, a UFO is a relative thing. The people in them know what they really are. The UFO sightings are only unidentified to people who don't understand what's happening.

I was at Edwards AFB until March of 1968 when I got out of the service. The Corona program is just a segment in a developing chain of events for reconnaissance programs. I don't know where the chain goes after Corona, but it goes to satellites. I'm sure there are space reconnaissance programs going on, but they don't need aerial recovery anymore. I don't know if Corona was a good thing or a bad thing, but it changed the course of history and helped end the Cold War.

> *"I don't know if Corona was a good thing or a bad thing, but it changed the course of history and helped end the Cold War."*

Chapter 4

Capt Donald R. Curtin

"The recovery program was very good and the aircrews were very efficient."

Capt Donald Curtin in 1967. /Photo credit: Donald Curtain

Capt Curtin (1932-) was interviewed on July 3, 2002. He was the first to be interviewed for this book. He was one of the original C-119J copilots assigned to the 6593d Test Squadron (Special) in 1958. Except for a 1963 to 1964 tour in Vietnam, he was continuously assigned to the aerial recovery mission from 1958 to 1966. During this time, Capt Curtin was also assigned to the 6594th Aerospace Test Wing, Operating Location Number 1 (OL-1), at Edwards Air Force Base (AFB) in California, as both a pilot and a copilot for C-119J and JC-130 aerial recovery flight tests and JC-130 checkout flights. He piloted the aerial recoveries of Ashcan balloon packages, and copiloted the JC-130 recoveries of Corona space capsules. Capt Curtin piloted an estimated 400 parachute recoveries while flying test and training flights.

Capt Curtin's journey began when he graduated flight school. He joined the 6593d in 1958 where he was introduced to satellite capsule recovery. He discusses how he was assigned to the 6593d and his thoughts on the Discoverer mission.

I graduated from flight school at Webb AFB, Texas, on November 1, 1955 and earned my wings. I was sent to Randolph AFB in Texas, and they put me in the multi-engine training down there in the C-119. Then in 1956 they sent me to Sewart AFB, Tennessee. I was assigned to a brand new organization, the 331st Troop Carrier Squadron flying the C-123. I was a copilot. They checked me out as an aircraft commander, because as soon as you had the flight time (at 500 hours) you were checked out if you qualified. I was there until 1958, at which time I was assigned to the 6593d Test Squadron (Special), a C-119 outfit at Edwards AFB. They were forming it to recover satellite capsules.

I thought it'd be fun. At that time, I was twenty-six years old. I know that every 1958 aircraft commander in the 6593d had flown in the Grand Union project back in 1954, I believe it was. They were qualified for parachute recovery. It had been a few years since they had done it, but they were all still very good.

Curtin started his aerial recovery flight training with aircraft commander Larry Shinnick. He explains the process of retrieving a parachute and the training methods used to teach him. Curtin also recalls squadron briefings about the Discoverer program.

I was a copilot and a first lieutenant. Now, at my experience level, even though I was a fully qualified aircraft commander, I was still just a copilot as far as the 6593d was concerned. I was assigned to an aircraft commander by the name of Capt Larry Shinnick. Larry was a super nice gentleman who had been a sergeant in the U.S. Army Air Forces during World War II, and had been wounded over Germany in a B-17. He stayed in the service and went through pilot training.

What we did in those days, Larry would make a parachute recovery, and then he'd say to me, "OK, now you recover one." This is how Larry and I did it. I don't remember what the other aircraft commanders did with their copilots. They weren't checking out any of us copilots in 1958 to be upgraded to aircraft commanders at that time.

We started doing practice recoveries at Edwards. There were two ways we did it. If you were doing a buddy system, you'd have another aircraft flying near yours. The other aircraft just dropped a practice capsule out above you, and then your C-119 lined up on the capsule parachute and recovered it. Or you could actually drop a practice capsule yourself and then recover it. We used to do that too. That was very doable.

We would go up to 15,000 or 16,000 feet and throw the capsule out. The guys put out the parachute recovery rig, and then we descended at about 1,500 feet a minute. We tried to keep our speed at between 120 and 130 knots, and came in to make the recovery. It was that simple. Those parachutes were special because they had reinforced bands around them, so the brass hooks from the airplane could dig into the banding and snag the parachute. That's how we did it. We just practiced. You practiced doing recoveries until you were good.

> *"We just practiced. You practiced doing recoveries until you were good."*

Normally, it took nobody more than two passes to recover a parachute. If you made two passes and missed an actual satellite capsule, there was another airplane behind you, and then it was their turn to try and recover it. The recovery program was very good and the aircrews were very efficient. You were assigned to the same crew and aircraft.

We reused our practice chutes. If they weren't torn too badly, we'd just repack the parachutes and use them again, because it was expensive. We used our practice parachutes over, and over, and over. In fact, my guys got pissed-off when you tore through a practice parachute that had been used many times.

They said Discoverer was strictly a scientific program up to Discoverer 13. It was not classified at first. When we first went to Hawaii, the newspaper over there had pictures of our squadron and the people. All of a sudden, they classified it, I think after Discoverer 14. Once Discoverer was classified, we were told nothing, absolutely zero, about its mission. Right after they classified it, they brought us all into a room and said, "From now on, you men can't talk about it anymore. You can't say another word about Discoverer." Then we could no longer talk about something that we'd been talking about for two years. That's the way the service is. Discoverer became a "black" program. The 6593d didn't call the program "Corona." They kept using the word Discoverer. I didn't even know it was called Corona until years later.

Chapter 4 - Capt Donald R. Curtin

An aerial recovery flight crew and their C-119J #18039 in 1958. Left to right (kneeling): 1st Lt Donald Curtin, 1st Lt Bobby Dorton, Capt Lawrence Shinnick, and the crew photographer; (standing) A1C Forrest Slaton, SSgt Aubrey Turner, TSgt Edward Hollifield, TSgt Stanley Sojda, SSgt Haynie Knight, and A2C Herbert Ponder./Photo credit: USAF

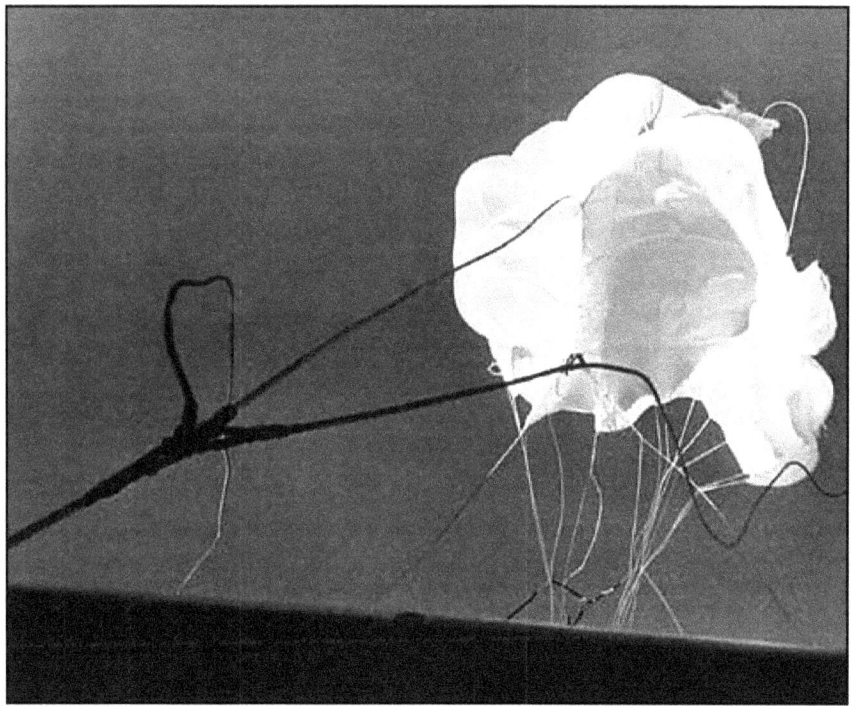

A parachute being recovered by a C-119J./Photo credit: USAF

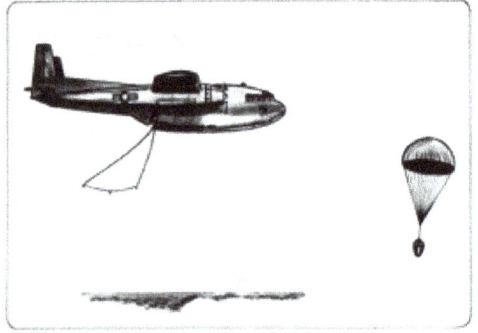 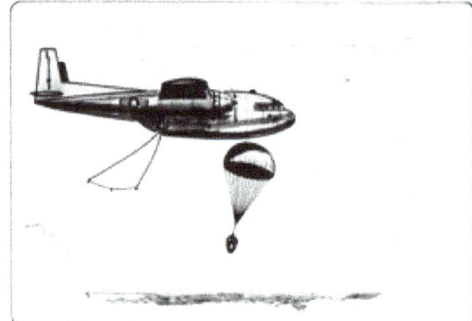

A midair parachute and capsule recovery by a C-119J./Photo credit: USAF

There was no way they could make the recovery flights secretive. They tried to, but we were taking off from Honolulu. You have to understand how Hickam is laid out. Hickam does not have an active runway. You have to taxi over to Honolulu and talk to the Honolulu tower. People knew we were taking off.

Most of the men were pretty good about keeping their mouths shut. What's there to talk about? One package looked just like another. Are you going to open the capsule and look at it? No, of course we weren't going to do that. We didn't care what was in it. All we cared about was doing our recovery job. Most of the pilots probably figured out Discoverer for themselves.

The squadron used two types of airplanes for the aerial recovery mission. It started with the C-119J and then transitioned to the JC-130. Curtin copiloted and piloted both of them. The planes had to go through renovations prior to becoming operational for recovery missions. He explains how they were acquired, maintained, and flown. Curtin later describes the changeover from the C-119 to the C-130.

The 6593d's C-119s were taken back from the Air Force Reserve. I think they were out of St. Augustine, Florida. The airplanes were in bad shape. They overhauled every C-119 we had, and put new overhauled engines on them. After the C-119s arrived at Edwards, we had to re-qualify them for satellite capsule recovery with All American Engineering Company equipment onboard. All American made the recovery equipment. Our people at Edwards reconditioned the airplanes to the point where they could be flown properly, and then we flew them over to Hickam AFB, Hawaii, and there they really went over them with a fine tooth comb.

"They were the greatest maintenance section in the world. We had the greatest maintenance officer, Capt Charles Hauenstein. He was just terrific."

They were the greatest maintenance section in the world. We had the greatest maintenance officer, Capt Charles Hauenstein. He was just terrific. Capt Hauenstein was allowed to transfer people from any base he wanted, and he went for people he knew in Tactical Air Command (TAC). He called crew chiefs and other people, and just grabbed them. They were the greatest guys in the world. They really knew their job. I never had an engine backfire on a mission. Those airplanes always ran like a jewel. Not one time did a C-119 ever abort a mission while I was there; I returned to Edwards in 1959. Every one of those airplanes flew.

We stayed at Edwards AFB until December of 1958 when we left for Hawaii. We left one crew and one C-119 (tail number 18041) behind at Edwards for flight testing. The man in charge of that crew was Capt Jack Parker and 1st Lt Floyd Barrow was the copilot.

We finally flew all the C-119s to Hawaii. Every one of those airplanes made it. In order to get the C-119s across the ocean from California to Hawaii, we were at Travis AFB, California, for three days waiting for favorable winds, because we didn't know if we could make it across the water. We had to have favorable winds in order to get those airplanes over there, because we were going westbound, so you get west winds. We had 1,000 gallons of 115/130 high-octane gasoline in the fuselage Benson tank. I couldn't get Larry Shinnick to burn a fuel tank dry in our airplane and then switch to another tank. He just burned them all low, but we were almost out of gas when we arrived. The C-130 could go from Hawaii to California with no problem.

You couldn't dump the fuel in a C-119, until they were later modified. The 1,000-gallon Benson tank was incapable of being dumped. If you had an accident, if you had an engine failure on takeoff or something, you were just a bomb. If fact, we had a waiver from the federal government. The C-119 was designed to fly at 69,000 pounds, but we had a waiver from the federal government to fly them at 74,000 pounds. Isn't that wonderful? We could fly them at 74,000 pounds.

The C-119 was also not able to be ditched. You can't ditch it, God's truth. It's a gull wing airplane, and the fuselage is not pressurized. If you try to ditch a C-119, it's going to sink right to the top of the airplane, just like that. Forget it. So, when we ferried those airplanes to Hawaii, it took us almost fourteen hours to get there. We had four hours in the middle where if we lost an engine, we couldn't go either way, because the airplane burns more fuel flying on one engine than it does on two.

At that time, the Coast Guard had ocean stations. Ocean Station November was halfway between Hawaii and California. You'd always talk to them on the way over and on the way back. They picked you up on radar. They were nice people.

When we ferried those C-119s out to Hawaii, they said our plan of attack was, "If for any reason you lose an engine, go to Ocean Station November and bail out." That was it. That was our plan to go to Hawaii. So, you went to Hawaii and said, "OK, keep your fingers crossed guys." That's the way it was in those days. After they were replaced by C-130s in Hawaii, the C-119s were ferried back to the U.S. and returned to the reserves where we had obtained them in 1958.

The C-119J Flying Boxcars assigned to the 6593d in 1959 at Hickam AFB./Photo provided by Charles Dorigan

When our squadron was formed, we had to use the C-119 because they were the ones flown during Grand Union. The C-119 wasn't a bad airplane. We had terrific maintenance and never lost a C-119. We wanted to transition to the C-130. The C-130 could go faster, higher, and further.

> "The C-130 is a great airplane. I loved that airplane. I have about 3,500 hours of flight time in it. The C-130 was a very, very excellent platform for the job that we were using it for."

The C-130 could take off with a maximum fuel load with no problem. In fact, we used to always go with max fuel loads. No sweat. It had four engines. The C-130 is a great airplane. I loved that airplane. I have about 3,500 hours of flight time in it. The C-130 was a very, very excellent platform for the job that we were using it for. It was my very favorite airplane in the military.

We went to Marietta, Georgia, and picked up a C-130 from Lockheed. They also grabbed C-130Bs from other commands. I think it was Troop Carrier Command. The airplanes were given to us, because we had the highest priority in the United States at that time. We carried a letter that said, "Department of Defense, Priority Number One." They sent the airplanes back to Robins AFB in Georgia and Lockheed was involved. Lockheed went to Robins AFB and said, "OK, we need these airplanes modified."

In 1959, Capt Ed Mosher accepted a job to go back to Edwards AFB and qualify the C-130 for capsule recovery. We needed a bigger recovery airplane with more speed, etc. Ed came to me and asked, "Do you want to go with me?" I said, "I'd love to." When we got there, we were on temporary duty (TDY). Then we became a permanent, detached unit, OL-1 of the 6594th Aerospace Test Wing. The unit included Ed Mosher, Jack Parker, Floyd Barrow, two navigators, about twenty enlisted personnel, and me. Because we were a detached unit, I never had one additional duty nor did my enlisted personnel, never once. It was great. We had a C-119, a C-130A and a C-130B. I got checked out as an aircraft commander in 1959 at Edwards, not in Hawaii.

I went back to Edwards for four years. I loved it. It was the greatest four years of my life. I was single for many years there. To me, Edwards was just the most exciting time of my life. I had a chance to work on the satellite recovery project, develop the aircraft, and work with Lockheed and All American Engineering. It was super, just great.

We went into a flight test program at Edwards. We were running tests all the time, constantly. We went to a briefing and Lockheed told us what they wanted. We laid out the flight test with our men and the people involved, set it up, and then we'd go do it. If it required Edwards, we'd contact the people who ran the range, telemetry, radar, etc. It continued that way the four years I was based there, and I am sure it was the same for as long as it continued operating at Edwards.

Most of our flight testing was done at Edwards, depending on the project we were running. Sometimes we were at other areas, but it all depended on what the test was. We worked out of different places: the Salton Sea Navy Test Range, California, the Point Mugu Naval Air Weapons Station, California, the White Sands Missile Range, New Mexico, and the China Lake Naval Air Weapons Station, California.

Originally, we were qualifying a C-130A #53-3131, which had been used for testing at Edwards AFB. When we got the airplane, it had a lot of hours on it, and it had been much abused. Number 53-3131 was a good airplane. We were using the same recovery gear in the C-130A that the C-119s had out in Hawaii. It was an upgrade, because a C-130 was faster and could pick up heavier capsules.

Aircrew members of OL-1 with their C-130A at Edwards AFB about 1960. Left to right: (kneeling) Capt Edward Mosher, 1st Lt Donald Curtin and TSgt Frank Kenyon (standing) TSgt Billy Hendon, A1C Walter Johnson, A1C Charles Dorigan, SSgt Lawrence Bradley, and SSgt William Culpepper.
/Photo credit: USAF

JC-130B during a parachute recovery flight test at Edwards AFB, circa 1960./Photo credit: USAF

We went through C-130B models. The JC-130B was capable of picking up a 3,000-pound capsule. We did satellite recovery testing with 30-pound capsules or less to 3,000-pound capsules with the JC-130B models. I flew the C-130A, the C-130B, the C-130E, and the C-130H. I flew four of them. I loved that airplane. To this day, I would love to fly a C-130.

> *"I flew the C-130A, the C-130B, the C-130E, and the C-130H. I flew four of them. I loved that airplane. To this day, I would love to fly a C-130."*

They totally modified the C-130s. They put a whole new capsule recovery hydraulic system into them, because we had to go up to a 3,000-pound capsule. The system had to be completely redesigned and All American built that one too. All American was good at what they did. They did a good job of designing and installing the hydraulics into the JC-130. We put extra hydraulic pumps on every engine and a whole new system in the C-130. The original C-130 had four hydraulic pumps, one in each engine. The JC-130 had four additional pumps to run the parachute recovery dolly. You had two separate hydraulic systems. When they went to that, the winch was now capable of doing so many things. They had to design it completely different with a stainless steel trough that ran down the middle of the dolly.

The copilot had four switches by his side for the recovery hydraulic system. Unless we were going to do a recovery, you turned the switches off to depressurize the recovery hydraulic system. When you were going into the recovery mode, you snapped those switches on. Now you had eight hydraulic pumps running, the ones for the aircraft's systems and the ones for the recovery system. That's the way it operated.

While I was at Edwards, we ran every recovery aircraft that went to Hawaii through a complete test program for satellite recovery, except the JC-130Hs. They were tested while I was in Vietnam. We brought the JC-130 back to Edwards AFB and kept it for about a month to run it through. We checked every system in the aircraft to make sure it had no problems. We kept them for awhile at Edwards, so we could make sure there were no leaks in the systems, make sure the recovery gear worked, and make sure the ECM gear worked properly. We kept the airplanes there and then we ferried them out to Hawaii.

The aircrews were also tasked to recover telemetry capsules from ships. The process of recovering a capsule was more detailed than merely snatching a parachute in midair and reeling it in. Curtin explains the roles and functions that all worked together to retrieve a capsule.

When they started doing Discoverer launches, a telemetry ship sat on the equator to track the Vandenberg launches. One of our missions out of Edwards was to pick up the data capsule off the ship and return it to Vandenberg. We used to do that constantly out of Edwards. The C-119 at Edwards flew down to North Island Naval Air Station, California to refuel. From there you'd go down just about to the equator, and make a pickup off the fantail of the ship.

The capsule was attached to a recovery line that went between the ship and its sea anchor. The rear of the ship had a pole that stuck up about 30 feet in the air, and there was a sea anchor about 100 feet behind the ship. A recovery line with about five loops in it was stretched between the pole on the ship and the sea anchor. The capsule was on the ship, next to the pole, and it was attached to this recovery line. The capsule was very light.

The way we were originally doing it, we flew in about 30 feet off the water with a single recovery pole hanging below the left side of the C-119. The recovery pole had a hook on it. With just a single line on the capsule, we came in with our radio altimeter set for 30 feet. The recovery line was out far enough that we were 50 feet away from the back of the ship when we went over it. With the back end of the ship going up and down, you'd try to adjust the aircraft. The fun part was trying to judge your up and down motion with the ship's up and down motion. You were pushing and pulling on the yoke. You hooked into the ship's recovery line with the airplane pole, snatched the capsule with the telemetry data in it, and pulled it onboard the airplane. Once in awhile the wing tip might go over the deck, but we were still way above the deck, so it didn't scare anybody. They were used to us. It was kind of fun. Later on, we used the C-130, which was much better. It had a much longer range.

They revised the system and the ships started using a balloon. Then we used our full recovery rig. The balloon had banding straps around it, and it was reinforced to take the full impact of the recovery system, the same as a parachute. The balloon was raised about 35 to 40 feet above the ship and we came in. That was easy. Now all you were doing was just hitting the balloon, going right through it, and snatching the capsule right off the back of the ship. That was it.

In about 1962, they had a telemetry ship north of Hawaii for some strange reason. At that time, it was part of Corona. They sent me out of Edwards with JC-130B #57-0526. The ship had some special information they wanted, and the ship was maybe 150 miles, 200 miles from the Hawaiian Islands. I think the space capsule had already been recovered, because we were sent to Hawaii to strictly just get the data.

The weather was terrible. The operations officer, Maj Gene Jones, wanted me to go make the pickup, but the ship said the weather was crummy. The ceiling was less than 400 feet. I said, "I can't do it. Can you send one of your crews?" He wouldn't do that either. So the ship started towards California. We stayed around in Hawaii for the mission. So, the next day Hickam's recovery control unit tried to send us again, but the weather was still terrible, so I refused.

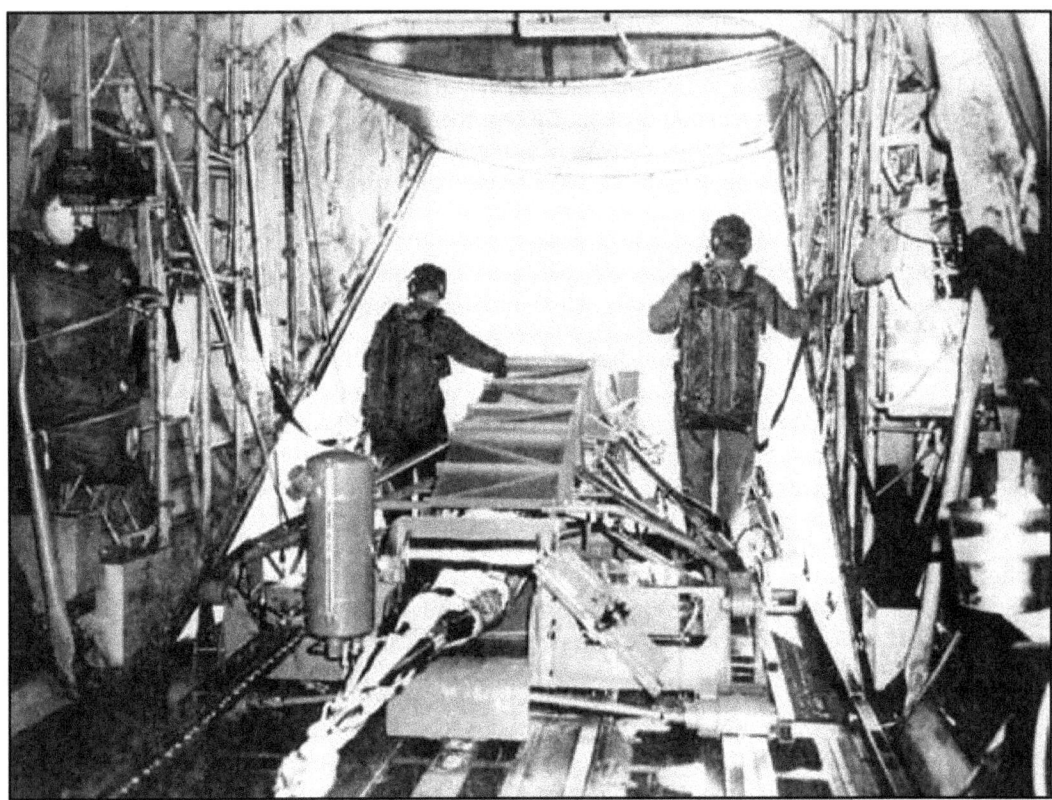
Two JC-130 loadmasters with the recovery dolly between them, circa 1966./Photo credit: USAF

We started back to Los Angeles. I had Capt Romain Fruge out of the Los Angeles Air Force Station (AFS) onboard as a passenger. They called us en route on high frequency radio out of the Recovery Control Group at Hickam. They said, "The ship reports the weather is clear. Would you go in and do a pickup?" By this time, we were halfway across the Pacific. We were 1,000 miles from each coast and the ship was about halfway between California and Hawaii. We said, "Sure. The weather is good." We went to the ship's frequency. We talked to the ship and they said, "Come on in and we'll put the telemetry data up for you." So, the ship put their sea anchor out, and we descended from 29,000 feet and made a data pickup. We took the package to either Sunnyvale or Vandenberg, dropped it off, and then we went home to Edwards. It was an interesting trip.

In the beginning, the ship data pickups were near Hawaii. As far as I know, I'm the only one who ever made a mid-ocean data pickup between Hawaii and the mainland. Edwards-based people were the only ones that ever made the ship data capsule pickups south of Vandenberg.

The difference between the C-119J and the JC-130 recovery equipment was the hydraulics system. The JC-130B dolly was driven by a hydraulic system installed at Robins AFB. The winch operator in the JC-130 drove the dolly back, hydraulically and electronically, and put it into position for a recovery. He had total control over it. We flew depressurized when we were making a recovery.

Prior to a recovery, the winch operator had to know the weight of the capsule. He had to prepare the recovery system ahead of time with the package weight in order to set the proper tension on the winch.

The winch operator adjusted the settings to the winch, so maybe 120 feet of line would payout. The dolly was capable of recovering 30 to 3,000-pound capsules. If the capsule weighed 150 pounds, the winch operator probably set it to 180 feet. The slack in the recovery line allowed the package to slow down, and then the winch operator just reeled the parachute in with the winch. The winch was set up like an inertial reel with a brake on it. If the winch operator set too high a tension on the winch, you tore through the parachute, because the parachute stopped too quickly after you snagged it. The parachute would have been gone. You also did not want the winch brakes set too loose, because you lost the line.

The pole handlers prepared the recovery gear for its operational mode. First they prepared the recovery rig while the 34-foot poles, the line and attached hooks were still inside the airplane. They attached the recovery pole hooks to the spring clip on their ends. The pole hooks had three prongs. The two hooks in the center of the recovery line had four prongs. Once the pole handlers put the hooks on the poles, the winch operator did the rest.

When the dolly was ready to be deployed, it was all done electronically and hydraulically by the winch operator. There were tracks bolted into the floor and the hydraulics drove the dolly along the tracks. The hydraulics system also lowered and raised the poles. The dolly went back and it went down. The winch operator could lower the poles about 40 degrees, hydraulically and electrically, while getting them into position.

The capsules had different weights. The C-119s worked with 30 to 200-pound capsules. The heaviest capsule I ever recovered weighed 3,000 pounds. We dropped the 3,000-pound capsule from an Edwards JC-130. I believe it was #53-3129. Then it was recovered with JC-130 #57-0526. We used an extraction

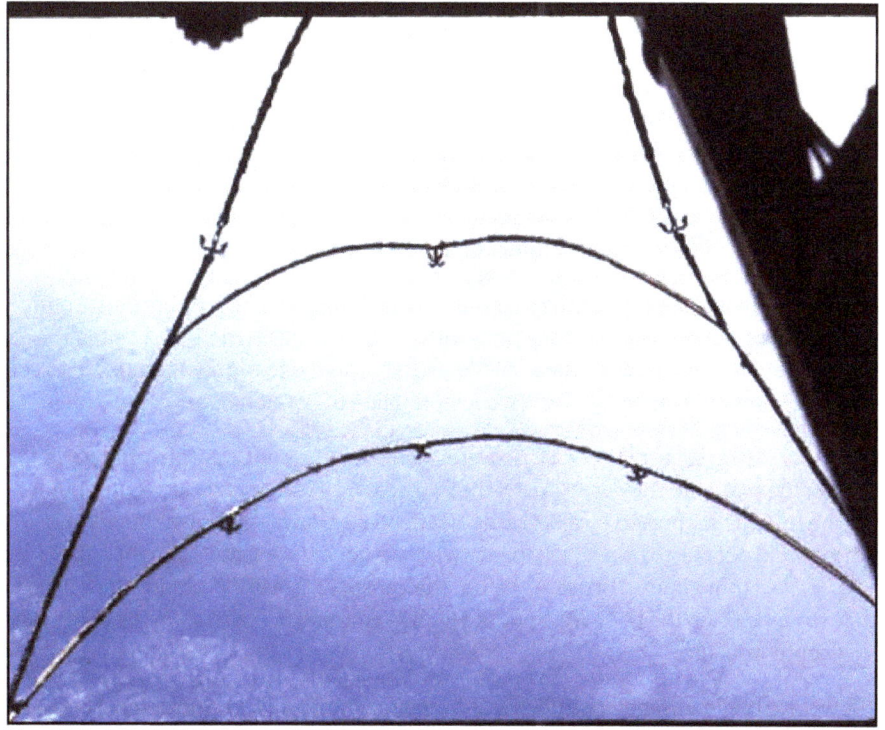

A parachute recovery rig (poles, line, and hooks) suspended below the rear of a JC-130 preparing to recover a parachute./Photo credit: USAF

parachute during those recoveries. This was normal in an aerial delivery system. They installed rollers on the cargo deck of the C-130, tied the package down, and when it was time to release the 3,000-pound capsule out of the airplane, they removed the restraints. The extraction parachute pulled the package from the back of the aircraft.

The parachute that held the capsule was designed to descend at about the same rate as the smaller ones. Every capsule, regardless of its weight, was designed to descend at about the same rate of speed, so the parachute was designed accordingly. You didn't hook into the main canopy when the JC-130 recovered it. The main canopy had to be very large to support a capsule that huge. There was a smaller 25-foot drogue canopy above the main parachute. Of course, it had reinforced cable all the way down to support 3,000 pounds, and we only hooked into the drogue on top.

The aircraft recovery line paid out like a fishing line. I'm not a fisherman, but if you throw out the line, it goes out and stops, and then you use your reel. You hooked into the parachute, and when the capsule got up to the dolly, they raised the poles and brought the whole dolly inside. That was it. We'd close the ramp and cargo door, re-pressurize the airplane, and zoom for home. It was very simple.

There was a fluid process to each mission. Curtin discusses what happened during a successful mission, from the initial briefings to the capsule's delivery. In addition to the plane's crew, there were several other entities that played essential roles in the Discoverer program. It was very much a team effort.

When there was a Discoverer/Corona launch out of Vandenberg AFB California, our mission control went on alert at our group headquarters. Somebody in the flight operations office (which I did later and enjoyed thoroughly) was given the primary control and they went down to our ops room. You were constantly pumped information from the Satellite Control Facility at Sunnyvale, California. You watched a map and kept track of where the satellite was, and you'd be told which orbit they intended to bring it in on.

Then you'd get your orders. They told the C-119 aircrews what time to start their engines and get in position, what time they were supposed to take off, and where they were supposed to go. They just said, "Here's the project. Synchronize your watches. Here's where the capsule is expected to come in. Go to your stations. Take off." Originally, we sent eight C-119s when they first started the Discoverer recoveries. They later sent nine after they acquired another one.

Sunnyvale tried to maneuver the capsules so they'd come out of the atmosphere just south of Hawaii. Sometimes they came out north, but mostly just south of Hawaii. That's where we wanted the package to be. There was less aircraft traffic south of Hawaii. When they wanted to trigger the capsule to come out of space, Sunnyvale had to signal the Agena/Discoverer satellite the orbit before.

When they started doing some of the first Discoverer launches, they sent a U-2 out of Edwards to Hawaii to provide data for the program. They used one of the two-seater U-2s that had Electronic Countermeasures (ECM) gear. The U-2 was supported by a B-47. The U-2 flew over Hawaii at high altitudes to help track the package. The U-2 did not talk to us recovery pilots; we were not on the same frequency. The U-2 collected data for Sunnyvale. From about Discoverer 16 on, the capsule recovery program started becoming so efficient they stopped using the U-2s. We also used RC-121 Warning Star aircraft in the early launches, and we had a helicopter squadron.

The C-119s were in predetermined positions and we waited until we picked up a signal from the capsule. The Discoverer packages had a beacon onboard. The capsule started sending its signal right after the

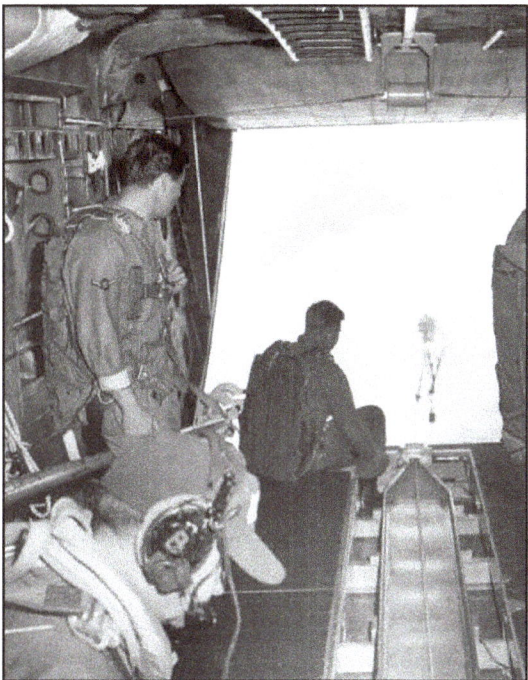

C-119J loadmasters recovering a training parachute and payload in 1959. Left to right: MSgt William Ramsey, A3C Donald Brown (standing), and SSgt Lawrence Bradley./Photo credit: USAF

Discoverer 26 being recovered in 1961 by a C-119 flown by Capt Jack Wilson./Photo credit: USAF

Agena ejected it. The capsule had a signal on it that even the C-119s had the capability of picking up. Our navigator could pick it up once it started transmitting, and we had a special Aerial Direction Finder (ADF) up in the cockpit. He could put the signal on there and a needle pointed directly to where the package was. The minute the C-119s picked up the signal, two or three of them headed for the capsule.

If the aircrew was sharp, they could see the parachutes at 40,000 to 45,000 feet. You could see the package coming in, because you knew where to look. You knew where the ADF needle was pointing. You could see the parachutes with no problem. The parachutes were orange and white. The capsule recovery parachute wasn't a regular parachute like a person jumps out of an airplane with. It had extra banding straps, 6 inches apart, all the way up to the canopy, and they were reinforced fabric.

The C-119s recovered the capsule when it got down to 15,000 to 16,000 feet. If a parachute was in front of you, the weather was good, and the capsule was doing well, your recovery was just a matter of lining up the parachute. It was like taxiing an aircraft. You put the airplane so your nose wheel is going down the center of the taxiway. In the left seat, the centerline is by your right leg. It was the same thing with catching a parachute. The only difference is it helped to have a horizon. You put the top of the parachute on the horizon, and you just kept descending at 1,500 feet a minute. It was up to your copilot to keep you informed of your altitude, airspeed, and your rate of descent. That's all you wanted to hear. You were descending nose-down, and you were not looking at anything except the parachute.

The C-119 recovery poles were just a little more apart than the width of the canopy, which was approximately 25 feet wide. So, when you hooked into the parachute with a pole hook or a line hook, you'd catch one of the straps. You'd tear through the parachute silk of course, but you'd hook into the strap. The winch payed out the line and picked up the package. Payout is the amount of rope that you wanted the winch to release after the parachute was snared by the rope.

Then the parachute was reeled onboard, and they closed the back end of the airplane. The parachute was never removed from the capsule. They wanted to see the complete package. Everything was put into a steel drum and it was sealed. We then took the capsule back to Hickam. It was a very good system and was that simple. It wasn't very difficult to do. If you had a good horizon, you had no problem, whatsoever. It was very easy, extremely simple.

The primary recovery aircraft assignment was rotated every mission. They went through the squadron and took turns. There would be a primary aircraft and a secondary aircraft. If the package came in where it was supposed to, the primary aircraft had the priority to recover it. He was allowed two passes. If he didn't catch the parachute after the second pass, the backup airplane could make approaches.

The Discoverer 13 capsule splashed down in the Pacific Ocean on August 11, 1960 and was recovered by a Navy diver and helicopter. After the capsule was brought to Hawaii, Curtin and his C-130 aircrew flew the capsule from Hickam AFB to Washington, D.C. Curtain describes the flight when he transported the historic Discoverer 13 capsule.

Quite often we were given a mission out of Edwards to go to Hickam and be there waiting when the Discoverer/Corona package arrived. We'd load the capsule aboard our airplane, and take off for Sunnyvale. That was the way it was done. I did a lot of flying back and forth to Hawaii, many, many, many times.

The first satellite capsule, Discoverer 13, was recovered from orbit on August 11, 1960. That capsule was recovered in the sea by a ship, because the capsule came in a little bit long and the recovery aircraft couldn't catch up with it. Col Charles "Moose" Mathison went over to the ship in a helicopter to recover

Col Charles Mathison (squatting) with the Discoverer 13 capsule (within the metal container) upon its arrival to Hickam AFB in a Navy helicopter on August 12, 1960. /Photo credit: USAF

Col Mathison (left) and Capt Edward Mosher carrying Discoverer 13 to a C-130A that would fly it from Hickam AFB on August 12, 1960./Photo credit: USAF

Capt Curtin was the copilot of this C-130A that transported Discoverer 13 from Hickam AFB to Washington, D.C. /Photo credit: USAF

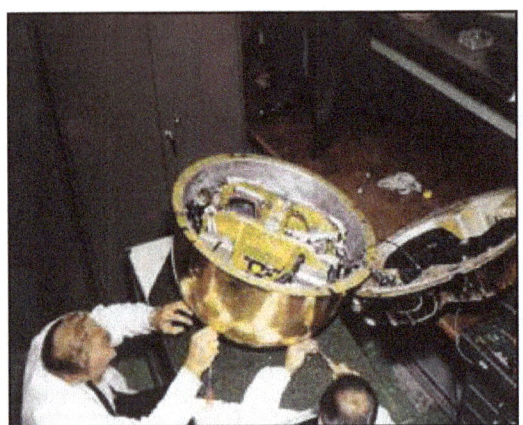

The Discoverer 13 capsule being opened at Sunnyvale. /Photo credit: USAF

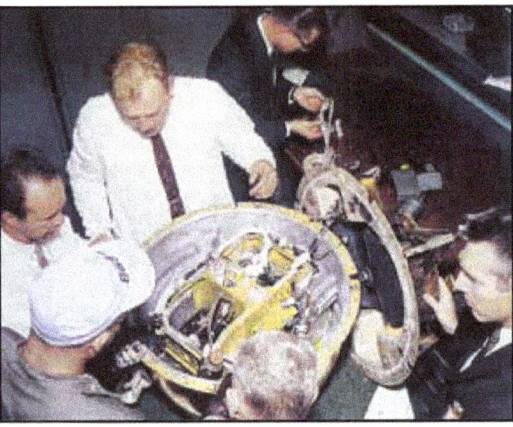

Col Mathison (wearing the hat) is pictured in this photo of the Discoverer 13 capsule being opened./Photo credit: USAF

The Discoverer 13 capsule when it was delivered to Andrews AFB in a C-130A copiloted by Capt Curtin on August 13, 1960, left to right: Lt Gen Bernard Schriever, Gen Thomas White, and Col Mathison./Photo credit: USAF

Chapter 4 - Capt Donald R. Curtin 121

The Discoverer 13 capsule at the White House on August 15, 1960, left to right: President Eisenhower, Gen White, and Col Mathison./Photo credit: USAF

the capsule. It is my understanding that Col Mathison actually unsnapped his 45-caliber pistol holster and threatened the captain of the ship to give him "his" capsule, which the captain did.[1] The capsule was in a drum-type container, and it was brought to Hickam AFB where it was loaded onboard our C-130A #53-3131.

I was the copilot flying for Capt Edward Mosher. Col Mathison arrived so we got our clearance and took off. We filed for San Francisco, but we knew we were going to divert to Moffett Naval Air Station near Sunnyvale. About halfway through the flight, SSgt Hansel Doug Stinnett came up to the cockpit and said, "Capt Mosher, you've got to see what's going on back there." Capt Mosher turned to me and said, "Don, go see what's going on." So, I went back to the rear of the airplane.

I went down to the bottom of the stairs. I looked at the back of the airplane and said, "Oh, my God." I got back upstairs to the cockpit just as quickly as I could. Col Mathison never saw me. I told the commander, "Ed, you will not believe what I've just seen. Col Mathison has the Discoverer parachute strung from one end of the airplane to the other. He's got the space capsule open, and he took the parts out of it." Ed said,

---

1   Several of the 6593d veterans and other veterans who participated in Discoverer 13 (met by the editor) heard that Col Mathison threatened the captain of the *Haiti Victory* in order to gain possession of the Discoverer 13 capsule. The editor did not find any official proof of this. In a 2008 interview, Col Mathison said the captain wanted his ship to deliver the capsule to the shore, but he relinquished the capsule, without being physically threatened, when Col Mathison insisted on flying it to Hickam AFB.

"Oh my God. What's going to happen when we get to Sunnyvale?" I said, "I don't even want to know, if it's alright with you." Just before we arrived at Sunnyvale, Col Mathison took everything and dumped it back into the container and resealed it. Nothing was put back where it was supposed to be.

Even though we had filed for San Francisco Airport, the media figured we were going directly to Moffett. When we landed, of course, flashbulbs were going off like crazy. The package was removed then taken over to Lockheed at Sunnyvale. It didn't involve us so much as it involved removing the capsule from the back of the aircraft. It was a very interesting time.

We sat there for two or three hours waiting for them to bring back the capsule to take to Andrews AFB, Maryland. When Lockheed saw the container, they did not know what had happened until they opened it up. They were just completely dumbfounded. Lockheed almost lost their collective minds and could not understand why the colonel did it. I do not blame them, because there was no reason for that container to be opened. It should have been left closed. Col Mathison was his own person, that's all. I'm not criticizing the man.

Anyway, we sat around for about two hours while Lockheed finally got their pieces out of the capsule. They gave us back the Discoverer capsule with a flag inside of it to take to Andrews AFB. This was all nonstop. We went for over forty-eight hours without getting any sleep. It was a long day, a long night, and we eventually got to Andrews.

As we were taxiing into Andrews, Col Mathison insisted on sitting in the left seat of the aircraft. The aircraft commander, Edward Mosher, told me to get out of my seat, and Ed took over the right seat. He told the colonel not to touch anything, and Ed taxied the airplane in and parked it. Col Mathison saluted Gen Schriever and gave him the capsule. There was an American flag inside of it supposedly, and it was presented to President Eisenhower.

We stayed a couple of days at Andrews, got some rest, and we went back to Edwards. It was a very interesting experience. What can I say? I didn't see the Discoverer capsule again until it was displayed in the Smithsonian Institute. I have seen it many times.

Curtin's C-130 flew south of Hawaii to provide top cover to the floating Discoverer 15 capsule. He briefly discusses their role in the mission and the unfortunate loss of the capsule.

When I was based at Edwards, sometimes we'd go over to Hawaii and fly satellite recovery missions. We qualified the C-130A #53-3131 to recover a capsule up to 200 pounds. Once we qualified it, they started sending us over to Hawaii to fly missions. We started flying the missions in the C-130, and they still had the C-119s. We used to fly top cover; they called it "downrange." We would be the furthest aircraft downrange, because we could go faster and higher than the C-119s, but we never made an actual Discoverer recovery.

I can still remember Discoverer 15 on September 15, 1960 like it was yesterday. The day before that mission, we'd taken off early in the morning from Travis AFB. During the mission, they told us that the capsule came in long. Discoverer 15 didn't land where it was supposed to. It was discovered by an air/sea rescue C-54 that picked up the beacon. The capsule was in the water down near Christmas Island. We went south in our C-130, because the C-119s couldn't make it. The C-54 was there when we arrived.

The commander of the C-54 happened to be the head of the air/sea rescue operations, a lieutenant colonel whose name I don't remember, who was talking over the radio to our Recovery Control Group commander, Lt Col Teuvo "Gus" Ahola. He asked Col Ahola if he could put his frogmen into the water and

Chapter 4 - Capt Donald R. Curtin

put the capsule in a rubber raft. The capsule had a salt plug. I forget how long the capsule salt plug was good for, but we knew it was eventually going to sink. The commander was asking, "How long are my people going to be in the water?" We were going to put a top cover over them; in other words, an airplane would constantly be overhead while they were in the water. They had a ship that could have arrived in ten or eleven hours.

We were circling the capsule for at least four hours. We put about eight hours on the airplane, and we were getting a little short of fuel. We couldn't make it all the way back to Honolulu, so we had to go to Christmas Island to get some fuel. We went over there and got fuel and they fed us. They were very nice to us, very pleasant people. Unfortunately, the decision to deploy the air/sea rescue men into the water was not made in time to save the capsule. The capsule sank and it was gone forever.

Curtin remembers a few stories that demonstrate the challenges he and his crew faced. Be it day or night, in the rain, or equipment malfunction, the 6593d Squadron rose to the occasion. He proudly reflects on their efforts and teamwork with heartfelt gratitude and admiration.

It was very difficult to make a recovery at night. The darkness made it very difficult. They had to put a light on top of the parachute so you could see it. That parachute was extremely difficult to hit if you didn't have a moon or a horizon. We didn't do many night recoveries, fortunately. Thank God. I'm talking about recoveries for another program, not the Discoverer/Corona program.

Aerial recoveries in a rain shower were no problem. You still needed some kind of a horizon. Real heavy rain would be very difficult.

If a capsule landed in the water and somebody could get to it, they put the capsule on a raft with a 15-foot pole. We flew flight tests where you'd fly in with what they called a "bare rope." You'd set the radio altimeter on the airplane for 30 feet, which gave you just enough room for the aircraft poles to clear the water. You'd fly in, hit the raft hook with the recovery line on the capsule, and snatched the capsule. In this case, it was a practice capsule.

During a test in 1961, we were flying our assigned JC-130B #57-0526 out of Edwards AFB. We flew to the Salton Sea to see if we could pick this 2,000-pound capsule up out of the water. We could see it floating about 15 feet from the recovery hook on a raft. So, we set the radio altimeter to 30 feet, and I flew across the water at about 125 knots indicated. You had to rotate the airplane just before the raft hook disappeared under the nose, so you could be climbing as the recovery gear went through the hook.

Lockheed forgot to use a floatable rope between the capsule and the raft. The rope was about 200 feet long, about 2.5 inches thick, and it was attached to a 2,000-pound capsule underneath the water. When I recovered the hook, the capsule immediately sank because it followed the rope underwater. The capsule went down to the bottom of Salton Sea, and we were trying to drag it out. The airspeed indication on our C-130 went from 125 down to 60 knots and quit; airspeed indicators quit at 60 knots.

I firewalled all the engines and pushed the throttles as far forward as they would go. As the airplane was being pulled into the Salton Sea, I was thinking, "Oh my God. We're going to hit the water. We're going in." In the back of this airplane, the dolly had a guillotine that was capable of cutting through the stainless steel cable we were using to pick up this capsule. That airplane is extremely noisy, but SSgt Bill Culpepper was standing back there and said he heard me screaming, "Cut it Willie. Cut it!" I was not talking on the interphone, because I had one hand on the yoke and the other hand on the throttles, and I was just screaming at the top of my lungs.

Divers in a water pick-up station preparing a recovery pole and a floating capsule for a flight test recovery in 1960.
/Photo credit: USAF

Bill was strapped against the right side of the fuselage near the winch. We always kept a man strapped to the wall of the aircraft with a battery switch in his hand for the guillotine. It was a very, very sharp blade, and that thing could cut right through the cable. I mean it could slice cable like it wasn't even there. Bill squeezed the switch and the guillotine came down. It cut the cable, left the capsule behind us, and we took off like a shot! We immediately accelerated. I was trying to hang on to the airplane, and eventually I regained control of it. God bless Willie Culpepper. He's dead now. He died of lung cancer.

The guillotine sat just above the trough at the very back end of it, near the winch. It was mounted in one of the trough's sections and it raised up about 12 to 15 inches above it. The guillotine was only armed during missions, of course. If you needed the guillotine, all you had to do was hit the switch and *BANGO!* The guillotine dropped 12 inches and *WHAMO!* It went right through whatever was down there. It had enough force behind it. The guillotine had a pyrotechnic charge that just drove it straight down. It cut through anything. We used it a couple of times in testing. If you couldn't bring the capsule onboard, you had to get rid of it, so you had to be able to cut it. I never heard about the guillotine being used at Hickam, because they never used heavy packages while I was there.

We went back to Edwards AFB, and my crew wanted to kill the Lockheed people. I bought them two cases of beer and I said, "Please don't do it. I would like to kill them myself, but don't do it." We had many interesting incidents, but that was very interesting because our people knew better. They knew that rope had to be floatable, and somehow we didn't see it as we flew over the capsule. I don't know why we missed it, but we did. It was never in the briefing for the test program.

Sometimes the enlisted men in our unit at Edwards challenged our officers to ride the dolly during a practice recovery when we were testing the JC-130Bs. I don't remember how many accepted the challenge, besides me. Maybe none of the officers were as crazy as I was. You put on a parachute, you got on the dolly while it was inside, and then it was extended out the back of the aircraft. You rode it out to the recovery position outside of the aircraft while we were in flight. You only had your hands to hold on with. No belts! You stayed out there for several minutes. It was fun, and I proved to the enlisted guys that I was as crazy as they were. I wasn't worried about riding the dolly because I had my parachute on. We never rode the dolly without a chute. That would have been a little stupid, even for us.

I think that the people at Edwards did more to help the program over in Hawaii than anybody. I will always feel that way. I knew some of the men that were based at Edwards who took over as our commanders, Maj Jack R. Wilson was one, and Maj Ed Bayer. They knew what they were doing. They were given a job and they did it. We did one heck of a job as far as I'm concerned. I think we did a terrific job.

C-119J recovering a floating capsule during a 1960 flight test./Photo credit: USAF

Chapter 5

SSgt Charles J. Dorigan

A1C Charles "Chuck" Dorigan in 1961.
/Photo credit: USAF

"The riggers working at the very rear of the aircraft were secured to the aircraft by a web 'dog leash' to keep them from accidentally falling out of the plane."

SSgt Dorigan (1938-) was interviewed using e-mail between June 2003 and June 2005. Dorigan was a twenty-year old enlisted loadmaster when he became one of the original members of the 6593d Test Squadron (Special) in September 1958. He was assigned to the 6593d until October 1959. He was then assigned to Operating Location Number 1 (OL-1) at Edwards Air Force Base (AFB) from 1959 to 1964 as a C-119J and JC-130 loadmaster for aerial recovery flight tests.

Dorigan was taught how to perform aerial recovery through on-the-job training, as most of his fellow loadmasters did at the 6593d. Fortunately, they had several veterans of the Drag Net project to learn from.

In September 1957, I reported to the 2d Aerial Port Squadron (APS) at Sewart AFB. At Sewart most of my flying was on C-123s during air cargo flights, or on troop drops, and on practice assault landings with the 82d and 101st Airborne Divisions at Fort Bragg and Fort Campbell. None of us young loadmasters at Sewart were trained in aerial recovery techniques before being assigned to the 6593d, but some of the older loadmasters at the 2d APS had been involved in the earlier Drag Net Project. Everything we learned about how to operate the recovery gear we learned on the job at Edwards AFB after we got there.

The Drag Net Project took place from late 1955 to early 1956 and involved C-119s that flew out of Japan and Alaska. These crews caught parachute packages released from balloons that had been launched from Europe and Turkey and had drifted across Russia and China while they took reconnaissance photos.

There were a number of older loadmasters at the 2d APS who had been on the earlier Drag Net Project. They wouldn't say anything about Drag Net, but prior to leaving Sewart for Edwards, they told me and my young colleagues that we would be doing a different kind of flying and there wouldn't be any loadmastering. They were correct. I never again loaded cargo or figured another weight and balance for the rest of my time in the Air Force.

I'd say just about a quarter of the 6593d enlisted crew members and quite a few of the pilots had flown on Drag Net. Most of the winch operators had flown on Drag Net, and I believe most crews had at least one loadmaster who had been on that early project. I looked through the first squadron history, the *History of the 6593rd Test Squadron (Special) (ARDC) 1 August—31 December 1958*, which contained a list of the original crews, and came up with fifteen people—crew engineers, winch operators and loadmasters—I know for sure were on Drag Net. There were several others I think might have been on Drag Net, but I am not positive. I arrived at Edwards in September of 1958 as an airman second class.

Most of the crew did not know what was in the capsules. However, they did understand the importance of the capsules and what they contributed to the Discoverer program. Dorigan describes his training and the crew assignments.

As I understand it, only the squadron commander knew what the capsules contained. All the rest of us (pilots, navigators and enlisted crew alike) thought the capsules contained mice, monkeys, or biological specimens. I have a number of Honolulu newspaper articles from that time, one of which is from the June 3, 1959, *Honolulu Star Bulletin* states, "Mice Cone Orbit in Doubt. How to Snare Capsule—First Photos."

The sequence of events from the time I first got to Edwards, to my learning about catching satellite space capsules, had a lot to do with how I perceived what we were doing on the project. Basically, it was flying recoveries first and learning about satellites afterward.

I started flying on training recoveries shortly after I arrived at Edwards, so I was involved in catching things in the air for quite some time before I thought about satellites. Also, recoveries had been made on Drag Net. Someone had done it before, so the process didn't seem unlikely. In fact, because of Drag Net, I felt somewhat like a newcomer to a previously established operation. We really didn't know for sure what we were going to be catching, and I do not remember anyone indicating we'd be catching balloons, along the lines of Drag Net, or anything else.

In September of 1958, it had only been about seven or eight months since the United States had launched its first satellite, Explorer 1, and I made no connection whatsoever relating that launch to the fact that we'd someday be recovering space capsules. All I knew was that I was part of this very secret mission and was engaged in some very different kind of flying that set us apart from other squadrons. It was kind of neat throwing things out of airplanes, reeling them back in and seeing the world from that big open back door. Not to mention being at Edwards and taking in all that was new in airplanes.

I simply can't remember when I learned that we might be catching space capsules. We used to have commander calls to inform us about things, but I can't recall whether we were told about capsules at these meetings, heard about them by word of mouth, or found out about them through the newspaper articles that began to appear in 1959. The first article that ever appeared in public about our squadron "Netting Capsules" was printed by the *Honolulu Advertiser* on January 8, 1959. I kept the first page of the article that shows our C-119s parked on the ramp and has a picture of the original squadron patch.

In 1958, the 6593d loadmasters were mostly taught by the ex-Drag Netters in space capsule recovery, and to some extent, by the tech reps from Lockheed and the All American Engineering Company. We'd learn about some new development in recovery methods from the tech reps either directly or by information passed down to us.

The squadron history states that All American Engineering representatives presented complete courses in the theory and techniques of the operation to all the tow reel operators (winch operators) and loadmasters, and that everyone received certificates of proficiency. The winch operators received that

Chapter 5 - SSgt Charles J. Dorigan

C-119J aerial capsule recovery./Photo credit: USAF

training because they were the key people on the recovery crew and needed to know how the gear worked. Personally, I do not remember any formal training. What I recall is learning my job by on-the-job training. I remember being introduced to each facet of the operation by the older crewmembers that showed us (hands-on) how to rig the poles, how to get the recovered packages back into the aircraft, how make recovery loops and maintain the equipment.

Early on in the project, we flew each day on different aircraft with different crews. Shortly afterward, permanent crews were formed and assigned to specific aircraft. Again referring to the squadron history, this was done to ensure maximum efficiency and promote flying safety. The history goes on to state that forming permanent crews expedited training, increased efficiency and established esprit-de-corps. It worked for us on the recovery crews.

I was assigned to Capt Warren Schensted's crew, transferred to Hawaii with him, and flew in his crew until I rotated back to Edwards in October of 1959. In 2002, I hosted the squadron's seventh reunion celebrating the forty-fourth anniversary of the unit's formation. Warren and his wife attended, and my wife and I sat with them at the reunion dinner. It was a pretty special moment for me.

For each mission, real or practice, we had to make sure that the winch was in working order, wound with good rope and that we had good poles and recovery loops. For the actual Corona missions, the aircraft carried a container (that looked like a 55-gallon drum) to store the capsule after a recovery. We picked up the parachutes, life vests and life rafts from the Personal Equipment Shop and delivered them to the aircraft. At Edwards, we often spent time at the Base Parachute Shop packing our own drop chutes.

I need to emphasize that the squadron made infinitely more practice recoveries than real ones—literally thousands of them over the life of the project. That didn't mean we recovered all the packages. It means we rigged and got ready for recovery, caught some parachutes, and missed others. The practice regimen kept up after our transfer to Hawaii, and in fact, our crew, with Capt Warren Schensted as aircraft commander, made the first practice recovery in Hawaii on December 12, 1958. I wrote down the names of the recovery crew on a torn piece of that parachute from that recovery, and at this moment that patch of parachute is on display at the NRO (National Reconnaissance Office) in Chantilly, Virginia. Two years and nine months later, on September 14, 1961, Capt Schensted made the first Corona recovery using a C-130. This was Corona/Discoverer 30, nicknamed "Twisted Braids" by the Air Force at Vandenberg AFB. Discoverer was the Air Force cover name for the CIA (Central Intelligence Agency) Corona project.

> *"I wrote down the names of the recovery crew on a torn piece of that parachute from that recovery, and at this moment that patch of parachute is on display at the NRO (National Reconnaissance Office) in Chantilly, Virginia."*

The first ever recovery of a film-carrying capsule, Discoverer 14, was made on August 19, 1960, nearly two years after the program began. After that, a number of rocket and recovery vehicle malfunctions occurred, two packages were recovered from the water and the C-119s only made four more live recoveries before they were retired in October of 1961. So, in between the time the squadron formed and the few real recoveries were made, the drill was to prepare the airplanes for recovery and practice, practice, practice.

> *"So, in between the time the squadron formed and the few real recoveries were made, the drill was to prepare the airplanes for recovery and practice, practice, practice."*

The entire loadmaster crew had to be proficient with the recovery equipment. There was some trial and error while they determined the most efficient methods of aerial recovery. Dorigan explains each piece of equipment and its function.

The primary C-119 recovery equipment we worked with every day consisted of the winch, the poles, and the recovery loop. Much of the recovery system was built on a metal frame covered with plywood, which was in turn covered by a black non-skid material. This formed the floor or deck of the recovery system. A shallow metal trough through which the winch line passed ran the length of the floor from the winch to the cable-cutter at the very rear of the deck. The trough was covered by a trough cover that was closed prior to making a recovery and opened while reeling in a package. When closed, the trough cover formed part of the deck and we could walk around on top of it. The trough cover was supposed to protect the crew by containing a flailing winch line, should the line break or have to be cut during a recovery.

The deck frame also supported two hydraulically-operated pole actuators and the rear pole roller-type supports. The pole actuators located at the rear of the aircraft raised and lowered the poles. The rear pole supports and the rollers supported the heavy handle end of the poles as the loadmasters pushed them in and out during the initial part of loop deployment. Handling the poles is how we got tagged as "pole handlers."

Chapter 5 - SSgt Charles J. Dorigan 131

Fixed to the top of the cargo compartment where the ramp opened at the rear of the aircraft was a roller device we called a "sheave." The sheave was used to help pull the package up pulley-style over the back of the aircraft. Pulling the capsule into the aircraft as it dangled below the rear of the airplane from the floor level was quite difficult, so being able to hoist the package upward over the deck was a great help.

Most of the cargo space between the winch and forward cargo bulkhead was occupied by two 500-gallon Benson fuel tanks covered by a box-like structure. The top of the box was used to stow parachutes and life rafts, to sack out, or to play cards. There was a pinochle craze going on at that time, and four crew members would sit on top of the Bensons and play the card game to while away the hours on some of our longer flights. On long-range missions, the full Bensons (forward of the aircraft's center of gravity) and the weight of our recovery gear caused us to be over gross and nose-heavy. This exception to fly out of limits, weight and balance-wise, was approved by the Air Force for those times when we needed it. The story is that if we lost an engine on takeoff with full Bensons, we would probably go down, although one pilot did lose an engine on takeoff and managed to bring the aircraft around.

I will try to point out or refer to the equipment described above in the photos that follow.

The winch was electrically-driven, mechanically-braked, and wound with 300 to 500 feet of half-inch nylon rope (see the photo below). The winches were designed to recover weights ranging from 80 to 300 pounds. When setting the winch for a recovery, the winch operator set the pawl, brake and delay. The pawl locked the drum in position and kept the winch from "creeping" prior to a recovery. The delay allowed the drum to turn so many turns before the brake began to engage. The brake could be set to account for capsule weight. The winch operator set the brake and delay settings in combination on the

MSgt Willie Stanberry, winch operator and recovery section Noncommissioned Officer in Charge (NCOIC), holds a piece of winch rope and the winch control box, while TSgt James Cross and All American Engineering technical representative Harry Conway look on. The winch control box was on a long electrical cord so the winch operator could move around the recovery deck with it. Behind MSgt Stanberry's head is a rear pole support. The large box-like structure behind the men covers the two 500-gallon Benson fuel tanks./Photo credit: USAF

winch to compensate for the package weight and to control how much winch line payed out. The idea was to get as smooth a payout as possible so there would be fewer G-forces on the packages and less chance to tear through a parachute or pay out all of the line.

The metal poles were 34 feet long and weighed about 150 pounds each. They were made of three metal tubular sections welded together. Early on in the project, there were two sets of holes drilled in the top side of the poles about 2 or 3 feet from the end of the pole in which were inserted a kind of spring clip. The spring clips were small wire cotter pin-looking devices in which we inserted the loop rope, and then tied a piece of string across the top of the clip to help hold the rope in place when the loop poles were down and rigged.

The early loops were made of 78 feet of half-inch nylon rope with three bronze hooks spliced into them. Two of the hooks were spliced into the loop so they could be fastened to the poles, and the third was spliced into the center of the "trapeze" portion of the loop that trailed between the poles. Later we stopped splicing the pole hooks directly into the loop and began fastening them to the loop with clevis pins so we could remove them more easily. We eventually covered the center part of the loop with a canvas sheath to prevent the nylon rope from burning through when it contacted the recovery parachutes. Sometime after, a second center trapeze loop was added that flew a few feet above the lower loop to improve our chances of snagging a chute. (The photo below shows a loop "in trail.")

One of the ways the engineers came up with trying to reduce the initial contact G-forces on the packages was to weave the rope back and forth along the length of the trough by tying the rope to small metal pins

The Discoverer 17 parachute and capsule on November 14, 1960. The loop is deployed and in trail. In this photo it appears that the chute has been hit and torn by the right pole, but the hooks remained taped to the poles./Photo credit: USAF

Chapter 5 - SSgt Charles J. Dorigan

A C-119 loadmaster taping a recovery hook to a pole./Photo credit: USAF

welded every few inches apart along the length of the trough. We tied the rope bends to the pins with a piece of stout twine we called break cord. I can't remember whether it was nylon or cotton. Needless to say, it took some time to do this, and we really groused about it at first. We got pretty quick at it, though, and could get re-rigged, ties and all, fast enough to make whatever number of recoveries were required during a practice flight.

We loadmasters spent a lot of time splicing loops down at the "section," as we called our place in the hangar. The half-inch nylon rope could be braided pretty tightly and was difficult to splice at first. The early loops with one center hook required eight splice points and the double loops required twelve splice points. My father had been career Navy, and I guarantee you I spliced more rope than he did. Early on in the project, I spliced so much rope that my fingers were sore. However, the task was made more amenable because of an ingenious little splicing tool that made splicing rope much easier than using the old marlinspike-like device. I've forgotten a lot of details about what we did in our work, but I've never forgotten how to splice rope.

The loops and the way the hooks were fastened to the poles were continually modified to improve their capability to more effectively snag the chutes. In the early part of the project, the four-tined hooks were taped to the poles with masking tape. An additional 1 or 2 feet of the loop was snapped into the two spring clips, cord was tied around the end of the spring clips to help hold the rope in them, and then tape was wrapped around the rope just before the rear spring clip to help keep the loop on during deployment and when the poles were down. It was learn by trial and error: too little tape and the hooks came off during deployment, too many wraps of tape and the hooks wouldn't pull off properly on contact with the chute. Eventually, a bayonet snap-clip modification was developed that allowed three-tined, claw-shaped hooks to easily be snapped onto the poles, yet held the loop in place and detached easily on contact with the parachute. The center hooks were always four-tined hooks.

The evolution of the hook fastening procedures was in essence a constant struggle to find a way to keep the hooks from coming off the poles during deployment, yet allowing them to break free on parachute contact with the fewest G-forces and without tearing through the parachutes.

Several tasks presented challenges to the loadmasters and winch operators. Dorigan describes typical duties and how the crew worked through them in order to have a successful recovery.

Prior to making a recovery, the winch operator set the brake and delay on the winch and the loadmasters fixed the hooks to the poles for eventual pole deployment. As mentioned above, on the early C-119 rigs we taped the hooks onto the poles with masking tape. We sat astride the pole actuators and leaned forward (really rearward out the back) to reach the end of the pole (see the previous photo and the following photo). The pole actuators extended a bit beyond the rear edge of the cargo floor, so you sat on the actuator and concentrated on taping this bronze hook onto the end of this metal pole, while at the same time you could see past the hook straight to the ground moving by several hundred to several thousand

Rigging the poles as demonstrated by the crew of aircraft #18043. Winch Operator TSgt James Cross (standing). Left pole: SSgt Thomas (Whitney) Mills rigging, A2C Norvel Jackson at the rear. The right pole is being operated by SSgts Herman Calling and Thomas Phillips. What is missing from this photo is a great deal of noise, a lot of slipstream, and Whitey's hair whipping around in the wind. The sheave is the yellow roller at the top of the photo./Photo credit: USAF

feet below you. There was a lot of noise from both the engines and from the slipstream whipping through the open beavertail doors.

As a former Tactical Air Command loadmaster, I was accustomed to working around open doors and ramps on troop and equipment drops, but the first several rigs did take some getting used to. You had to get the shank of the hook firmly against the pole and make sure the tape was wrapped very tightly and did not allow any movement of the hook, otherwise the tape would tear and cause the hook to come free. I'd often get very fumble-fingered trying to get the hooks taped on and the strings tied, and it was at this time I knew that open-heart surgery was not in my future. It was even worse when it was cold, as it often was in the winter at Edwards or at higher altitudes.

We had to straddle the actuators while rigging back then, because the poles rigged with the hooks would not fit through the actuators. The pole actuators were subsequently modified so they could be opened up, allowing the poles to be rigged outside of the actuators and inside the aircraft with the doors closed. This made things a lot faster and easier and we could rig long before we needed to actually deploy the loop.

When deploying the loops, the loadmasters at the handle end of the poles pushed the pole out as the two loadmasters at the hook end of the poles fed out the loop. Feeding out the loop took some practice at first. The hooks were buffeted around by the slipstream and were frequently thrown back into the aircraft, whipping the loop into a Gordian knot of hooks and rope that had to be laboriously untangled. It was especially annoying when the hooks came off the poles after they were lowered to recovery position or immediately prior to a contact, forcing us to re-rig again from the start. This was frustrating for both the recovery crews and the pilots, who were waiting to make another recovery attempt.

I became aggravated whenever the rigging failed, but I also became physically beat from slipstream and engine noise, the constant wrestling with the recovery gear, and horsing around the practice packages and capsules as we worked at altitudes up to 14,000 feet. The aircraft had oxygen hoses along the side of the cargo compartment, and we did take whiffs of oxygen between tasks at altitude. The hoses had little spring loaded caps at the breathing end and you'd flip open the cap and take a whiff of oxygen by sucking on the end of the hose. I can't remember whether we kept track of who breathed out of which hose, but sanitation wasn't the governing factor in getting the job done. I used to plug my oxygen mask into a hose because it was more effective that way. Often the pilots wanted to make more practice recoveries than we had planned, and I can remember returning from flights quite tired, and on some occasions with mild headaches from working at altitude.

In the beginning of the project we dropped packages from a designated drop ship. Flying drops was a good duty. We didn't have to hassle with rigging and recoveries, we'd just sit in the rear of the aircraft throwing out packages and taking in the great scenery from our lofty view. Soon, however, the crews became proficient enough at rigging to rig for recovery and drop packages to themselves.

What we'd do for a "self drop" is get the poles rigged and ready to deploy, then one of the loadmasters sat in the center of the floor at the rear of the aircraft and pushed out the practice package. The chute was out in a second or two and we'd quickly unhook the static line, toss the empty parachute bag out of the way and begin to deploy the poles as the pilot started his descent. According to what I read, early parachutes descended at about 33 feet per second or 2,000 feet per minute, which, counting parachute opening time, only gave the pilots about six minutes to make a recovery before the packages hit the ground. This was quite fast considering that on missed passes they'd have to fly outbound past the chute, then turn around and line up on it again. With flaps extended and power reduced, we'd come down fairly fast, but I also remember pilots cycling the landing gear to gain added drag in an effort to get us down fast enough to counter the high rate of descent of the early chutes. Later systems, as I read, descended at about 20 feet per second or 1,200 per minute, which gave the pilots more time to line up on the package. I can't remember for sure the most number of passes we made before catching a chute—perhaps four or five—but I do remember snatching chutes that were about to descend into the tops of the Joshua trees at Edwards or into the whitecaps off the coast of Oahu.

Sometimes we caught a chute without damaging it much. If the pilots wanted to try another recovery and there were no new chutes, we'd just stuff the relatively undamaged chute back into a parachute bag, hook it up to a package and drop it again.

I always enjoyed the moment we popped open the beavertail door prior to a recovery. The cargo compartment of the C-119 was dark inside, and after lumbering off the ground and reaching recovery altitude, we'd hit the door actuator and there'd be a blast of slipstream noise and this sliver of light that grew larger until we saw the world in an ever-moving frame through the big square opening in the back of the aircraft. After we rigged and started to make our pass at the package, we'd hear the power adjustments and feel the movements of the aircraft as the pilot lined up on the package. These movements were exaggerated to us in the back of the aircraft, particularly as we got closer to contact. Once in a while the pilot would really kick in a hard right or left rudder to try and snag a chute passing just outside the poles. These movements became less exaggerated as the pilots became more proficient.

In September of 1958, our first month of flying, we recovered 50 percent of our packages—34 out of 68 dropped. In the first six months of 1959, we pulled in 237 out of 252 packages dropped for a 94 percent recovery average. Three months in that period showed a 100 percent recovery rate. Though the pilots did miss some packages, many of the missed recoveries occurred because of winch or hook malfunctions, hooks tearing through parachutes or because the parachutes oscillated too much to be recovered.

If the pilot made a good center hook contact between the poles there was kind of a *whump*, then the loud whirring of the winch line paying out. If he made a pole contact, then there was more of a *thump* or even a *wham*, depending how high on the pole the contact was. We bent poles, broke poles, even had pieces of broken pole hit the aircraft. I kept a flight log and it shows that on September 24, 1958, we recovered two of three packages on #042 (C-119 airplane number 18042), but we hit the chute of the third package with the number-two prop and had to shut down the engine.

Some aerial recoveries encountered problems. Dorigan describes what some of these issues were and how the loadmasters reacted to them.

Sometimes we had problems reeling in a package after we hooked a chute and we had to cut the winch line with a cable-cutter. The cable-cutter was fired to cut the winch line for a number of reasons, two of the most common being to get rid of uncontrollable packages or to prevent winch over-speeds. Packages most often became uncontrollable because of the aerodynamics of the package or billowing chutes, causing the chute and package to whip around or jerk so much that we couldn't reel it in safely. Packages or hooks flailing around this way might also hit and damage the aircraft. The practice packages we caught weighed from 80 to about 180 pounds. They were made of concrete blocks, pieces of railroad track, Jet Assisted Take Off (JATO) bottles filled with shot, and some were simulated capsules.

Crew hoisting a capsule into the aircraft in 1959. The crew is using the overhead sheave roller to pull the capsule upward and over the back of the deck./Photo credit: USAF

Winch over-speeds occurred because of brake failures, incorrectly set brake and delay settings or the parachute being caught too low. If a pilot caught a chute in the shroud lines below the canopy, the chute would lop over and be pulled behind the aircraft like a drag chute. This pulled the line off the winch so fast that the over-speeding winch could possibly fly apart, spraying cargo compartment with shrapnel.

Assuming we made a good recovery and we didn't tear through a chute or have an inverted "drag chute" pull off all of our winch line, we then started reeling in the package. On the C-119s and the C-130A, the cable-cutter at the rear of the aircraft acted like a stabilizing mechanism for keeping the winch line reeling in smoothly. Once the package was reeled in to where the winch line was hooked into the loop with the clevis-type device, we had to open the cable-cutter to get the rest of the loop and entangled chute into the aircraft. Sometimes the package would continue to reel in quite smoothly after the cable-cutter was up, but at other times, the slipstream, the chute and the type of package caused the whole system to whip around in circles at the back of the aircraft, or thrash back and forth across the floor, or even to cause line payout. Working at the very back of the aircraft trying to stabilize these chutes and packages with their embedded bundles of hooks might have been dangerous, particularly if a package suddenly tore free, but I don't think any of us gave it more than fleeting thought.

For water and ship recoveries, Dorigan notes that the pilots required great skill. The plane had to fly at low altitude in order to snag the package. Dorigan explains the process to recover a package from the water and a ship.

Some Discoverer capsules went into the water and were recovered by boat or helicopter, and one sank (Discoverer 15) while I was still flying on Hawaii missions. When I was in the 6593d at Hickam, we practiced for a downed capsule water recovery by putting the capsules in life rafts or in floatation gear and rigging them for water pickups (as shown in the photo below). To set up a water recovery, pararescue swimmers (PJs) and rafts were dropped into the water near the floating test capsule. The PJs put the capsule in a raft, or next to it in flotation gear, then erected a 10 or 12-foot high pole with a hook on the top end of it, and that became the recovery station.

The pilots approached the raft at about 30 to 40 feet altitude, which put the ends of the poles about twelve feet or so above the ocean surface. As he flew over the raft, the pilot rotated the plane sharply and began to climb. The center portion of the loop, which had no hook in it, engaged the hook on the end of the pole and the winch began to pay out. The climb out combined with the proper winch setting was supposed to prevent the capsule from hitting the water as the winch line payed out. If the pilot flew "through" the station, or if too much line payed out, or both happened, the package would hit the water and skip or try to submerge. The latter event often caused a payout that pulled all the line off of the winch or caused us to cut the winch line to prevent a winch over-speed.

One of us usually sat all the way in the rear of the aircraft next to the pole actuators and reported how high the poles were above the water during the approach to the raft and what happened at contact. It wasn't unusual for an aircraft to get too low and drag a pole in the water, de-rigging the poles, or on

C-119 water recovery from a life raft station in 1960./Photo credit: USAF

Crew members prepare to remove the canister from a recovered data capsule. The person on the right has his foot on the edge of the trough. Each one of those little holes in the trough had a pin on the other side of it to which we had to tie one of the bends in our accordion folded rope. Lots of ties! And we had to retie each time we rigged for another recovery./Photo credit: USAF

occasion breaking a pole. We felt a lot of vibration in the pole actuators as a pole hit the water at 125 knots, as the pilots certainly did as well.

At Edwards we flew a variation of the raft pickup where we recovered telemetry data from a ship stationed about 800 to 1,000 miles south of San Diego. We'd fly the C-119 down to North Island Naval Air Station (NAS), take on a maximum fuel load, then headed south toward the ship. We used the A model C-130 later on when it became operational. The round trip in the C-119 took about eight to ten hours chugging along at 180 knots at 10,000 feet or so, and around four and a half to six hours in the C-130 flying at 300 knots at 25,000 feet. It was nice to have those two extra engines over all that ocean.

On ship recoveries we flew with only the left pole down and rigged with a hook. The ship towed a nylon line about 200 feet long behind it from a pole erected on its fantail. The line had a sea anchor affixed to the end of the rope that was in the ocean to keep the line taught. A series of loops hung from the line between the ship's fantail and the sea anchor, and the pilot aimed for these loops in order to engage the hooks. Attached to the line was a somewhat bell-shaped fiberglass container (see the photo above) weighing probably 30 pounds or so. In the container was a canister with data tapes, and at times, mail from the ship's crew. We gave the letters to people in base operations in San Diego to mail when we delivered the data tapes. Like the raft pickups, the pilots flew the approach at about 40 feet altitude, rotating just before contact so that we'd be in a climb when the winch started its payout.

I always thought it took a lot of skill for these pilots to fly the raft and ship pickups. On the ship pickups you had large aircraft flying very close to a large ship and the pilots had to compensate for wind, sea state, the ship's forward motion as well as the up and down movement of the ship's stern to which the recovery line was attached. Some will tell you matter-of-factly that it wasn't all that bad. Eventually the pickups were done by the balloon recovery method where the line to the data package was attached to a loop around a balloon flying about 100 to 150 feet above the ship's fantail. This was a much safer and surer recovery system.

I always enjoyed flying these water and ship pickups. At 40 feet above the water you could actually smell the sea air. As the water rushed by under you at 125 knots, there was this great sensation of movement, as there always is in low level flying. Being way back in the aircraft, you could feel the subtlest of control movements as the pilot lined up on the raft station or ship line, then the ship whipped by and there was the quick elevator ride up at rotation, contact and climb out.

Dorigan had the honor of designing the first 6593d patch. This "team" attitude was evident throughout the squadron. He discusses the team and his transfer back to Edwards AFB in 1959.

In late 1958, people started thinking about a 6593d squadron emblem patch. I like to draw, so I fooled around with some ideas, got some feedback from Danny Hill and others, and came up with a blue eagle on a white background catching a red parachute. I gave the drawing to Capt Tom Hines, who in turn gave me something like five dollars and a three-day pass. Some patches were made and worn by several crew

(Left) The first design of the 6593d Test Squadron (Special) patch that squadron members wore from 1959-1960, and (right) the modified squadron emblem that was officially approved by the Air Force on July 14, 1961./Patches provided by Charles Dorigan

The crew of C-119 #18042 at Hickam AFB in 1959. Kneeling, left to right: A1C Ken Riding (Photographer), A1C John Lansberry (LM), Capt Warren Schensted (P) and 1st Lt Jack Ludwick (Nav). Standing left to right: TSgt Elbert Jenkins (CE), 1st Lt Robert Clifton (CP), SSgt Billy Anderson (LM), A2C Charles Dorigan (LM), TSgt Leonard Champion (WO), and SSgt Matias Aragon (LM).
/Photo credit: USAF

members. The first patch was never officially approved by the Air Force. The patch appeared in a January 1959 *Honolulu Advertiser* article about our squadron, the first such article about the unit. In September of 1960, the falcon and lighting design was submitted to the Air Force as the squadron's official emblem and subsequently approved. It is really a distinctive patch. The falcon is indeed the Air Force bird, and the colors of that bird along with the lightning bolt make for a much livelier patch than my original.

Our recovery section at Hickam really didn't have a normal pyramidal Noncommissioned Officer (NCO) chain of command as you had in other organizations. We had a recovery section NCOIC and nine winch operators, each one was responsible for the four loadmasters on his crew. Among the ten original recovery flight crews, including the one that remained at Edwards, the crew lists show we had thirty-one NCOs and only nineteen airmen. So we airman were outnumbered by NCOs nearly two to one, but we didn't feel overwhelmed because we operated as crews with everyone on each of the crews sharing tasks in general. Obviously, if loops were to be made or some detail needed attending to, either for their own crews or for the unit in general, the airman would be first in line to do them. However, the NCOs, as often as not, pitched in to get the job done. We were more like friends than anything else, and I do not remember anyone taking advantage of this to cause any dissention.

We had two master sergeants in the recovery section, MSgt Willie Stanberry, the section NCOIC and MSgt William Ramsey. Both were winch operators. Sgt Stanberry had the ranking recovery crew in the squadron consisting of himself, three tech sergeants, and a staff sergeant.

TSgt Leonard "Champ" Champion was the winch operator on our crew. The loadmasters were SSgt Andy Anderson and SSgt Matias Aragon, A1C John "Frog" Lansberry and me, an airman second class. We also had an aerial photographer, A1C Ken Riding, assigned to our crew. Ken flew with us fairly often and photographed the recoveries, how we did our work bringing the snagged packages into the aircraft, or for that matter, what happened when things didn't work out as they should have. The photographers were temporarily assigned to us from the 1365th Aerial Photo Squadron in Florida.

We had a good recovery crew. Everyone worked well together and Champ, an ex-Drag Netter, was really the best of people. I believe he was one of the winch operators who came from the flight engineer ranks. He was very quiet and capable and had a good sense of humor. I think Andy and Matias were loadmasters by trade. Frog, however, was one of two people in the squadron who came from the target-towing career field. Prior to transferring to the '93d, he and SSgt Fred Stebbins flew in the back seats of T-33s and B-57s operating the tow reels used to tow targets for gunnery practice in Florida.

Capt Warren Schensted was our aircraft commander. He was then and still is a quiet, very gentlemanly person. He occasionally let me get some stick time from the copilot's seat because he knew I wanted to fly someday. Lt Bob Clifton and Lt Jack Ludwick, copilot and navigator, respectively, were also nice people. TSgt Elbert Jenkins, our flight engineer didn't say much and seemed to me to be kind of gruff, but he really wasn't.

Many of the recovery crewmembers in the 6593d, particularly the older ones, had followed some interesting career paths on the way to the Corona project. A good number were former aerial gunners who cross-trained to the loadmaster career field after the B-29s, B-50s and other bombers they had flown on were retired from service. Several had served as aerial gunners on B-29s in the Korean War. Winch operators Louis Bannick and Jim Muehlberger were combat veterans (airborne) of World War II, and TSgt Emory Head had been a prisoner of war of the Japanese in World War II.

All in all, we recovery crews had it pretty good flying on the C-119s in the 6593d at Hickam. We earned flight pay and got to do the type of flying that offered us stunning views of the earth. Though we had some squadron duties and some rudimentary barracks cleanup chores, we did not have to participate in any base details or pull kitchen patrol. That made a big difference in our quality of life. We were having a good time as far as Air Force assignments go. In October of 1959, I rotated back to Edwards with the crew that was to begin testing the C-130s for aerial recovery.

Upon his arrival at Edwards, Dorigan and other recovery crew members started flying on the C-130. Dorigan and his Edwards' aircrew flew flight tests and recoveries while making the transition from the old C-119 to the more advanced C-130 recovery airplane. Dorigan discusses his move and some difference of the 6594th Aerospace Test Wing at OL-1.

A recovery crew commanded by Capt Jack O. Parker was already operating there. When the 6593d staged to Hickam in December of 1958, Parker's crew had remained behind to test new recovery methods to be used on the C-119s. Our newly arriving crew was led by Capt Edward H. Mosher with 1st Lt Donald R. Curtin as copilot. The C-119 pilots had to fly check rides to transition to the C-130, but the recovery crews from the two units could work together right away because the C-130A model winch and pole rig were nearly identical to the C-119 rig.

Between the two crews, our detachment consisted of about five officers and fifteen enlisted men. People rotated in and out of the unit over time, but the complement at OL-1 was always around twenty to twenty-two people. Our group occupied two rooms in the Aerospace Test Pilot School hangar. One large

Chapter 5 - SSgt Charles J. Dorigan

The members of the 6594th Aerospace Test Wing at OL-1 wore this unofficial unit patch. This emblem was never submitted for official approval./Patch provided by Charles Dorigan

C-130A #53-3131 with its rig deployed in 1959./Photo credit: USAF

room was for the recovery crew and the other somewhat smaller room next to it was for the pilots and navigators. So, the officers and enlisted men were in much closer proximity on the ground than we were at Hickam where the officers and enlisted crews had ready rooms in different locations in the hangar.

The plane we came to Edwards to fly, C-130A #53-3131, was the second oldest production model C-130 flying in the Air Force at that time. It was one of the few old C-130s that still had the "grasshopper nose" that housed the old APS 42 radar antenna (see the bottom photo on the previous page). The majority of the newer A models sported the characteristic bulbous nose housing the new radar antennae. Edwards also had two of these old birds they used for various airlift chores, including carrying the fire trucks that might have to be used if the X-15 rocket plane made a forced landing on one of the dry lakes along its flight path. The 6511th Parachute Test Squadron at El Centro NAS also flew two old C-130s. What we noticed about #131 right off was that the pole actuators opened up and the hooks clipped to the poles instead of being taped to them. This made rigging much faster and easier. These modifications were eventually applied to the C-119s at Hickam.

Our recovery chain of command and crew duties at Edwards were much the same as they were at Hickam. We maintained the recovery gear, put the personal equipment on the aircraft, packed drop parachutes, and on long flights we picked up the coffee and flight lunches for the crew. We often helped the flight engineer prepare the airplane for recovery missions or cross country flights and put the plane to bed after flights. On engine startup, one of the loadmasters, wearing a headset and in contact with the pilots, stood by the front of the aircraft and monitored the engines for leaks or mechanical problems and made sure no one on the ground tangled with a prop. When all four engines were running, the loadmaster climbed in the front crew door and pulled it closed so the bird could begin to taxi. (In the top photo on the next page the loadmaster, A1C Jim Stewart, is monitoring the engine start for #131 on its flight to Washington with the Discoverer 13 capsule.)

> *"To me, the C-119 had a number of endearing qualities. The great sound of the R-3350s cranking up into a chugging start, their rumpety rump as we taxied, and the thrumming roar as we took off were real airplane sounds..."*

To me, the C-119 had a number of endearing qualities. The great sound of the R-3350s cranking up into a chugging start, their *rumpety rump* as we taxied, and the thrumming roar as we took off were real airplane sounds (though the C-130s sound neat, too). The C-119 cockpit with all of its "greenhouse" windows and navigator's astrodome bubble, had a more open aspect to it than did the C-130 cockpit. I spent a lot of time peering out of the navigator's bubble during night flights. My Navy father taught me the major star constellations and I had a great interest in navigation.

Although the recovery rig was heavy, the C-130A flew significantly below the gross weight of a regularly loaded C-130, and as a result had a high rate of climb. I used to ask Mosher and Curtin to make "max effort" take offs whenever we were traveling somewhere because the bird could really climb out for a big airplane. They'd start rollout, build up a good bit of speed, and then haul back on the yoke and up we'd go. It was fun to see the ground drop away so quickly, and I'm sure it amazed people on the flight line who saw our rapid climb. The performance specs that I've been looking at show a 2,000 feet per minute (fpm) rate of climb for the C-130, but that is probably for a loaded aircraft under normal flying conditions. I'm sure that we momentarily topped 2,000 fpm climb rate by a significant figure.

Starting up #131 at Hickam AFB./Photo credit: USAF

Rear view of the C-130A with the rigging deployed in 1959./Photo credit: USAF

We flew quite a few different kinds of recovery missions with #131. We flew many light capsule recoveries, loop configuration tests, instrumented pole tests, balloon recoveries, and ship recoveries. Several times we flew back to All American Engineering Company's facility at Sussex County Airport in Georgetown, Delaware, for refitting and recovery tests.

In 1961 we flew balloon recoveries in elevated recovery station tests at Fort Bragg. I believe these tests were part of the program to test the human ground pickup systems that had been previously done by "goal post" pickups and earlier versions of balloon recoveries. In balloon recoveries, the hooks engaged the recovery loop and line supported by the balloon flying some 150 to 200 feet above the ground. On contact, the line from the balloon to the ground began to move laterally while stretching some, causing the energy to translate down the line in such a way that the package on the ground first moved upward before it moved forward. This was called the "hidden pulley effect" and it is certainly useful in getting a person up before he goes forward, as in getting out of a clearing surrounded by trees. Via Capt Mosher, our crew was given a letter of commendation from the All American Engineering Company for the Fort Bragg tests, and Mosher had that letter made a permanent part of each crew member's personnel record.

An aircrew from OL-1 flew to Hawaii to support the Discoverer recoveries. Dorigan was part of the Discoverer 13, 14, and 15 missions. He explains his role and the events during each mission.

I was never on an aircraft that recovered a real Corona capsule, but we flew #131 to Hawaii in support of the Discoverer 13, 14, and 15 missions. On the hot missions we flew at about 25,000 feet, already rigged and ready to go for the capsule if we had the chance. If we sighted a capsule, the procedure would be to start descending and depressurizing as we headed for the capsule, then start deploying the loop and poles at about 14,000 feet after we had depressurized and slowed down to about 125 knots.

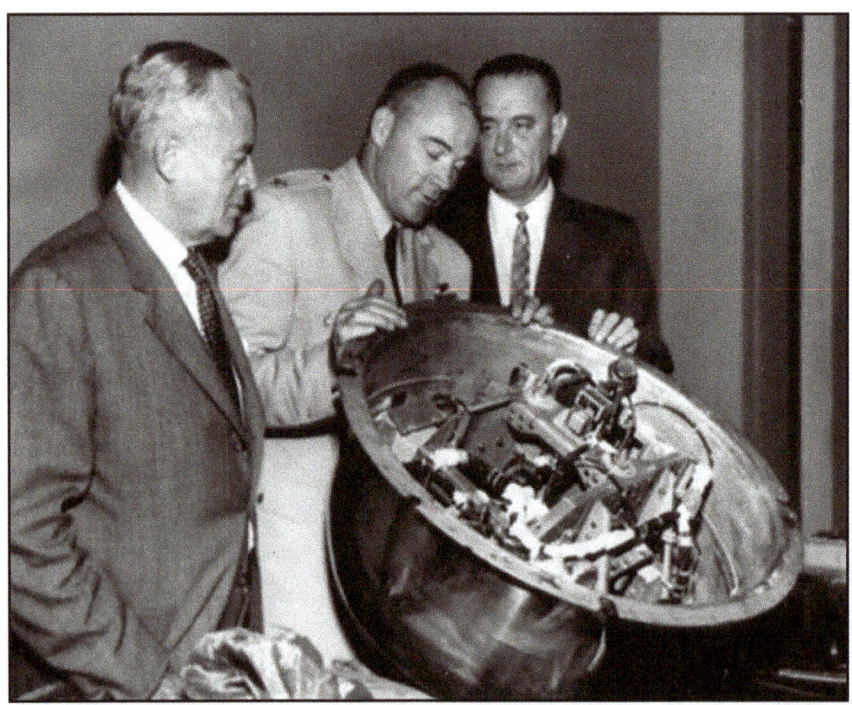

Senator Lyndon Johnson and Col Charles Mathison with the Discoverer 13 capsule./Photo credit: USAF

Discoverer 13 went into the water on August 11, 1960, and the Navy recovered it—the first object recovered from space. The capsule was brought out to #131, which flew it back to Washington, D.C. to be presented to President Dwight D. Eisenhower by Gen Thomas D. White, Chief of Staff of the Air Force. The flight crew consisted of Capt Mosher, Don Curtin, navigator Andy Radel, flight engineer Billie Hendon, winch operator Doug Stinnett and loadmasters Jim Stewart and Donald Hackworth. Col Charles "Moose" Mathison, commander of the 6594th Test Wing (the 6594th Recovery Control Group and the 6593d were subordinate units to the Test Wing), went along, and on the way back to Washington, he opened the capsule, much to the surprise of #131's crew, and apparently to everyone else in recovery operations. There was an American flag inside the capsule, and some think the flag was placed in the capsule on the ground after it had been recovered.

On August 19, Capt Harold Mitchell caught the Discoverer 14 film-carrying capsule. I don't think any of us realized at the time that we were part of an organization that was making reconnaissance history. The story is that only the 6593d squadron commander and some others up the chain of command knew what we were really catching. We thought we were catching roaches, mice, or monkeys. There was even a public relations cover piece in the press showing researchers placing a monkey in a Discoverer capsule. The whole thing was highly classified and we were told to say nothing to anyone about what we did or how we caught the capsules. It was "need to know" and only we needed to know.

After C-119 #037 caught Discoverer 14, some "capsule people" at Hickam took the package somewhere, looked at it, and then brought it to our C-130 to fly back to Moffett NAS near Sunnyvale. I asked the technician who brought it to us, "What's in the capsule?" and he said, "I can't tell you." Then he asked me, "How do you guys catch those things?" and I said, "I can't tell you either." It was need to know at its finest! We flew the capsule back to Moffett NAS and were photographed by the press as we unloaded the capsule on the ramp.

> *"I asked the technician who brought it to us, 'What's in the capsule?' and he said, 'I can't tell you.' Then he asked me, 'How do you guys catch those things?' and I said, 'I can't tell you either.' It was need to know at its finest!"*

An armed courier officer always accompanied the capsules on the flights to Moffett. On our way back one night, the courier came to me with a cup with something in it and asked if he could have some water. He then carried the cup of water back to capsule container, put it on top of it and began stirring up the contents. I thought for sure he was going to feed the monkey, but he was only preparing to shave so he'd look presentable when he turned the capsule over to the intelligence people.

Discoverer 15 overshot the ballpark and landed in the ocean not far from Christmas Island. Apparently, a C-54 had located the capsule so we flew down in #131 to join them and to provide an air cover in case they jumped their pararescue men into the ocean in order to put the capsule in a raft. We saw the capsule in the water. We had these marker beacons with us and we opened the ramp and threw out one to mark the capsule's location. The beacon was in a bomb-shaped device about 4-feet long with a weighted nose and a beacon and strobe light on the back of the bomb. We stood at the rear of the ramp, tossed out the bomb, and watched as it fell away from us and splashed into the water. We saw it bobbing around in the water near the capsule. Unfortunately, the PJs (pararescue men) were not given permission to jump. The capsule had a saltwater plug designed to dissolve in twenty-four hours to keep the wrong people from getting the capsule in case it floated too long or out of the area. The plug worked and the capsule sank before any ships could reach it.

We had been flying for a long time, so we landed at Christmas Island to fuel up for the 1,200-mile flight back to Hickam. As I mentioned, I'd always been interested in Hawaii and the Pacific, so I enjoyed seeing Christmas Island and the Gilbert Islanders who were brought there to work on the copra plantations. British airmen stationed there helped us with the refueling. I remember saying to one of them, "Did you know Captain Cook discovered this island on Christmas Eve of 1777?" And he replied, "Yeah, the bastard!"

Dorigan was involved in the JC-130 flight tests for the recovery mission. Each of the JC-130 models had similar recovery equipment, yet some differences as well. He describes the equipment, how it worked, the technology, and some dangers. Dorigan provided most of the equipment photographs below.

JC-130B #70526 arrived at Edwards in 1960. The first of the B models to arrive at Edwards, it was the last C-130 to leave recovery operations in Hawaii in July of 1986 after twenty-six years of flying recoveries. The #526 kind of eclipsed good old #131, as #131 did the C-119s. In terms of amenities, the new airplane had a bunk at the rear of the cockpit instead of a wall full of circuit breakers as did #131, and #526 also had a small galley fitted for heated coffee jugs, as opposed to #131's thermos jugs in which the coffee seemed to cool rather quickly. The A model had 3750 horsepower (hp) turboprop engines swinging 15-foot diameter three-bladed props, where the JC-130B had new 4050 hp engines turning 13-foot diameter four-bladed props. The bottom line was that #526 was quieter and didn't vibrate as much. The new airplane had a totally new recovery system like none we'd seen before. It also had a much more sophisticated Direction Finding (DF) system housed in the radome-like structure on top of the fuselage.

JC-130B #70526 with poles down and loop in trail at Edwards AFB on April 17, 1961./Photo credit: USAF

Chapter 5 - SSgt Charles J. Dorigan

Looking back from winch operator's position, March 1961. The photo shows the winch operator's console, part of the winch and the dolly at the back of the aircraft./Photo credit: USAF

The DF system was operated from a large panel in the forward part of the cargo compartment.

The components of the B model C-130 recovery equipment consisted of the winch, the dolly, the poles and loops, and the direction finder operator's station. We used the term rig for the poles and loop, and we said we "rigged the poles," but we also used rig as a generic term to describe the whole system, much like a truck driver calls his truck a rig.

The winch had two drums. One was wound with half-inch steel cable for heavy recoveries (up to 3,000 pounds) and the other was wound with nylon rope for the light recoveries (30 to 300 pounds at that time). The larger winch was hydraulically operated and the small one was operated electrically with mechanical braking as on the C-119s and the A model C-130.

The dolly was the big yellow boom structure that moved back and forth in the B and H model C-130s. It consisted of the dolly framework, the trough, the pole actuators, the fairlead, the bomb racks, and the guillotine cable-cutter.

The trough was the long box-like structure that ran fore and aft along the top of the dolly. The winch line payed out through the trough during a recovery. Then the winch line, loop and parachute materials were reeled back in through the trough after the recovery was made. The top of the trough could be opened up so that we could get to the winch line or parachute materials in case of a problem.

The poles held the loops to which the recovery hooks were attached.

The capsule cradle was used to hold the large capsules as they were initially pulled up to the rear of the aircraft, then trundled into the cargo compartment. The cradle could be removed if we weren't making heavy recoveries.

The direction finder operator's station was between the forward bulkhead and the winch.

The dolly boom is in the raised position at the very rear of the aircraft. The foreshortening in the photo on the previous page creates the optical illusion of the trough standing straight up. The boom could be raised and lowered and was usually raised to help lift in heavy packages. During recoveries, the boom was in the lowered position, providing a straighter path for the rope or cable to pass through.

The square object in front of the dolly (upper left corner of the photo on the previous page) is the cable/chute-cutter guillotine. It cut through the steel cable or whatever parachute or webbing was in the cutter at the time. The cutters on these early rigs were fired shut by a .45-caliber blank housed in the cylinder on the right side of the cable-cutter. The toggle switch on the winch operator's console that actually fired the cutter was a "guarded switch" with a red cover over it. The winch operator had to flip up the switch

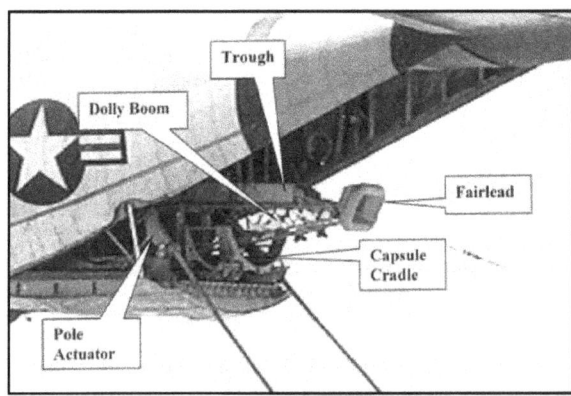

Close-up of the dolly all the way aft and in position for recovery.
/Photo credit: USAF

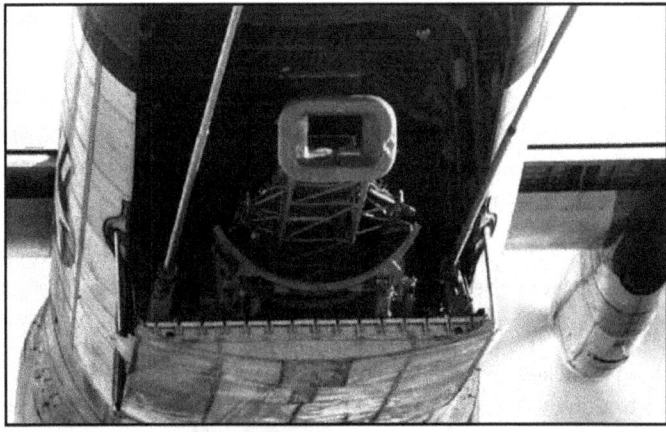

Aft view of the JC-130B recovery rig with poles up and unrigged at Edwards AFB on April 17, 1961./Photo credit: USAF

Chapter 5 - SSgt Charles J. Dorigan

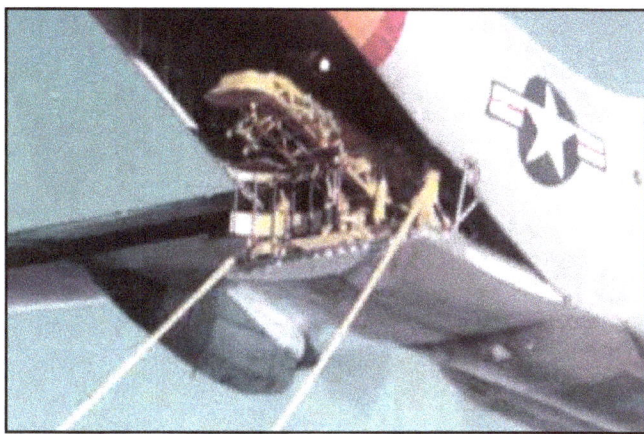

JC-130B dolly in the raised position with the capsule cradle tilted outward—poles down and unrigged. In this photo the fairlead is down and the capsule cradle or carriage is tilted outward. This is the configuration of the rig when a heavy package has been reeled up to the fairlead and is about to be set into the capsule cradle, Edwards AFB on April 17, 1961./Photo credit: USAF

cover before he could fire the cutter with the toggle switch. I believe the last item on the winch operator's checklist before a recovery was "cable-cutter armed." I do not think I ever put my hands or arms in that guillotine for any reason, even if the plane was just sitting on the ground with no one near the winch operator's console.

On the previous page in the bottom photo, the capsule cradle is in the horizontal or tilted back position. This is the position in which the cradle, containing the capsule, is eventually reeled back into the cargo compartment. The dolly is in the lowered position where it is when the rig is configured for recovery. Note that the fairlead has a big ding in it. This could have happened when a hook got hung up on the fairlead while the winch line continued to reel in line, causing the hook to bite into the metal of the fairlead. It could have also been caused by hooks or other hardware flailing around and hitting the fairlead.

The loadmasters working in the back of the aircraft helped stabilize the capsule so that it could be fitted to the cradle. Two of us would get back on the very edge of the ramp by the cradle and one of us passed a tie-down strap around the capsule to the other loadmaster. We'd cinch the strap tight so the capsule would stay in the cradle as the cradle was tilted back. Once the capsule was in the cradle, it was tilted back, and served as kind of a capsule holder/trolley that allowed the capsule to be reeled into the cargo compartment. The cradle could be removed so it wouldn't be in the way when working other kinds of recoveries or when dropping things.

The photo of a later model rig (see top photo on the next page) provides an excellent view of the whole recovery system. The direction finder operator's panel is in the left foreground in front of the crew member, who might be the DF operator. Then we see the drum-shaped large winch, the winch operator's console, and the recovery dolly rig with the two loadmasters on it all the way aft on the ramp. For reference: the cargo compartment is 40 feet long, 10 feet wide and has 4,500 cubic feet of useable space. The ramp, upon which the dolly is sitting, added another 10 feet to the length of the recovery deck. The DF operator seems to be looking down toward the winch operator, who might not be in his seat, but seems to be adjusting something on the winch. His hand and arm are barely visible in the lower right corner of the photo.

View of the whole recovery system./Photo credit: USAF

Deploying loop—about 1963./Photo credit: Charles Dorigan

Dolly aft, poles down and rigged./Photo credit: Charles Dorigan

Just prior to the actual recovery, the crew members usually positioned themselves on dolly at the rear of the aircraft or stood up front by the winch operator or DF station. Normally, no one stood on the cargo deck between the winch and the dolly during a recovery. This is the area where the rope or steel cable winch lines were exposed, and people standing in this area could be hit by recoiling steel cable or winch line if a line broke or had to be cut.

In the bottom photo on the previous page, A1C Frank James (left) and I are deploying a loop on the JC-130B. The capsule carriage has been removed for this light capsule recovery flight. The A-frame pole storage rack behind me was where we stored the end of the poles for take offs and landings or put the ends of the poles when we weren't using them. At this point in the deployment, the end of the pole is resting on the ramp with the hooks just over the edge of the ramp. As we deploy the loop, the winch operator moves the dolly aft toward the rear of the aircraft and the poles move outward and downward with it. The pole actuators are mounted on the dolly and ride back with it as it moves aft, and in a sense, take the place of the two aft C-119 and C-130A "pole pushers." Because of this, we could fly with a winch operator and only two loadmasters on many of our missions.

The dark linear object protruding from the side of the dolly is a "bomb rack" from which marker buoys were dropped.

When the dolly is all the way aft, the poles are lowered to their final down position. During the final part of the deployment, the loadmasters climb on the dolly and ride it back to the end of the ramp. As one rode the dolly back, there was a brief optical illusion that made it appear as if the dolly was going to roll off of the ramp. Sitting back there on the dolly always provided a great view of the countryside and the recoveries as they happened.

In the photo on this page, the dark bands on the center loops on either side of the center hooks are canvas sheaths that kept the nylon rope in the loops from burning through from the friction at chute contact. This particular loop has extra hooks spliced higher in the loop to snag the center of the canopy as it passes between the poles. The pole handle, the upright black object in the photo, was used to push the poles through the actuator. It fit into a notch on the actuator assembly and kept the pole from rotating in the holder. As I had mentioned before, the pole actuators could be opened so the poles could be lifted in and out of the open actuators, which made it possible to rig the poles quickly and with the ramp closed. But it was also handy for removing a pole that was bent or damaged in such a way that you couldn't pull the pole back through the actuator.

Even that didn't work at times. We once made a high hit on a pole, which bent the pole upward so badly that the end of the pole impaled itself in the aircraft's horizontal stabilizer. Because of the way the pole was bent, the force on the handle end of the pole in the actuator was so great that it made it impossible for us to remove the pins that let us open the actuator so we could remove the pole. Fortunately the end of the pole did not punch into the elevator. However, part of the bent pole extended below the open ramp, so Capt Mosher had to make a very flat approach to a landing on the dry lake to keep from dragging the bent part of the pole on the ground.

The practice package in the photo on the next page is an expended JATO bottle filled with the amount of shot it took to simulate a specific capsule weight, somewhere around 80 to 180 pounds. The funnel-shaped fairlead guided the hooks, line and recovered parachute as they were reeled into the trough. Here the fairlead is in the "up" position. The fairlead could be tilted down manually, or it tilted down automatically with the weight of a suspended heavy capsule.

The dolly and trough system on the JC-130Bs and JC-130Hs made reeling in light packages much easier and safer on those aircraft than on the C-119s and the C-130A. The troughs and movable dolly

View of the rigged poles, July 1964./Photo credit: USAF

Chapter 5 - SSgt Charles J. Dorigan 155

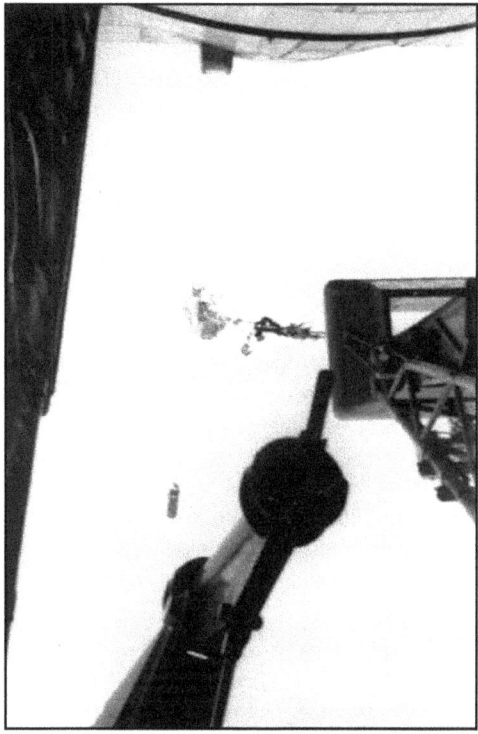

Practice package in trail, July 1964./Photo credit: USAF

kept the parachute materials contained and helped control the recovered package and materials as they were being winched in. The top of the trough could be opened in case the hooks or parachute material snagged on their way through the trough. The troughs prevented much of the thrashing around of the chutes, lines and hooks that often occurred while reeling in packages on the C-119s and the C-130A rigs. The large winch drum also allowed us to reel significant amounts of parachute onto the winch. This was important when hauling in heavy capsule suspension systems. They usually consisted of a recovery parachute with its suspension lines connected by a support-webbing to one or more very large load-carrying parachutes. Their suspension lines were in turn connected to the support-webbing connecting the parachute or parachutes to the capsule.

The photo on the next page is also a photo of a later model rig. I see several improvements over our old system, among them:

- A much larger and obviously more powerful cable/chute-cutter with the blade slicing downward instead of across the cutter opening;
- Rollers on the sides of the cable-cutter to prevent snags;
- A kind of biscuit mold flooring that was added to the dolly floor framework to facilitate walking around on the rig; and,
- Oxygen hoses that are attached oxygen regulators that seem to be located above crew members (foreground and standing on the rig) that allow more mobility than ours did. Our regulators were mounted on the walls to the side of us, and our hoses could get caught up in things or on the floor as we worked.

The riggers working at the very rear of the aircraft were secured to the aircraft by a web "dog leash" to keep them from accidentally falling out of the plane. Occasionally, we wondered what would happen if our chute popped and flew into the slipstream while we were still hooked to the plane. If that happened, the dog leash (rated at 5,000 pounds tensile strength) would hold us in the aircraft while our parachute canopy would try to pull us out of the aircraft with a drag force of something like 2,500 pounds, or so they said. That was not something we wanted to experience.

One day while I was working at the very back of the rig, my parachute D-ring got caught on the sonabuoy holder and my chute popped. Luckily, SSgt Willie Culpepper, who was right behind me, caught the pilot chute and I never did find out what it was like to become part of a drag chute system. The upshot of the whole thing was that Capt Mosher saw the incident as an opportunity to get our crews recognized by 6594th Test Wing Headquarters for our sometimes dangerous work. He recommended Willie and me for Wing NCO and Airman of the Month awards, respectively. We received the awards, but I had a bit of a time living that down, since I essentially received an award for messing up.

People had thought about the possibility of the loadmasters working without dog leashes, so that an inadvertently popped chute would not injure a person. We conducted tests in which we threw dummies out of the airplane to simulate a person accidentally falling out of the airplane while it was rigged for recovery. As I remember, the dummies did clear the rig, but the decision was made to continue to wear the dog leashes.

The recovery crew handles a recovered chute as it is winched in through the trough toward the winch./Photo credit: USAF

Chapter 5 - SSgt Charles J. Dorigan

The C-130 was flown for a variety of aerial recovery tests. Dorigan describes heavyweight recoveries, X-15 recoveries, long rope recoveries, and various other types of aircraft used in test recoveries. He also shares a story about a chase plane crash.

We flew quite a variety of missions using both the light and heavy recovery capabilities of the new rig. We flew many light recovery test missions that included aerial parachute reinforcement tests, recovering objects from the ground and water, testing balloon recovery systems, testing loop and hook configurations, and running pole tests using instrumented and Teflon-taped poles. The Teflon tape wrapped around the poles reduced the friction of the parachutes hitting the poles, which prevented pole breakage and allowed the chute to slip down the pole to engage the pole hooks.

On our heavy recoveries, we pulled in weights ranging from several hundred pounds to 3,000 pounds. We nearly pulled in a 3,500 pound capsule, but it caused aircraft control stability problems that forced us to drop the capsule. Like the light recoveries, our heavy recoveries also included aerial, ground and water pickups, some quite novel indeed.

Large capsule systems consisted of a 24 or 28-foot diameter recovery parachute, the suspension line or "web" connecting the recovery chute to the main support chutes, the main parachute or parachutes and their suspension lines and the suspended package. As I recall, one 100-foot diameter G-11 type parachute could support a 3,000-pound load for airdrop, and I believe we did at times use a parachute chute rig with the 3,000-pound capsules we recovered. Loads could also be suspended from multiple smaller chutes, such as the G-12s that were 64 feet in diameter and could each suspend a load of around 2,000 pounds or so. We might have used single G-12s for lighter practice capsules. The exact figures elude me after all of these years, but I think I am close. Again, the bottom line is that there was a whole bunch of parachute, parachute suspension line, and web connector strap to be reeled in before the capsule was ready to be secured to the cradle.

We dropped the heavy capsules to the recovery aircraft from our JC-130B airplanes. A heavy capsule was about 5 feet in diameter, and about 6 feet tall, and looked like a big stovepipe with a little stovepipe on top of it. I believe these might have been the simulated SAMOS capsules. We'd fasten roller conveyors to the floor of the aircraft, and load the capsules onto the aircraft with a forklift. We'd run the dolly forward, then roll the capsule into the aircraft on the conveyors, and secure it with tie-down chains.

We dropped the packages from higher altitudes, sometimes 20,000 to 25,000 feet. We'd climb to drop altitude, don our helmets and oxygen masks, depressurize, and open the ramp for drop. Just prior to drop, the loadmasters removed all of the tie-downs except the one that kept the capsule from rolling toward the ramp. About ten seconds out we'd remove that one. At the drop signal we'd push the capsule rearward as the pilot made a slight pull up and the capsule rolled out neatly. After it went out, we'd walk back to the ramp and pull in the static line and parachute bag. After pulling in the parachute bag, I always spent a little time on the ramp taking in the unparalleled view of the high desert, Lancaster, Palmdale, Mojave and the San Bernardino and San Gabriel Mountains that separated the high desert from the Los Angeles basin.

We tested ground and water recoveries of capsules of various weights and shapes. One that I especially remember was probably near 10 feet long, weighed about 800 pounds, and was shaped like a giant ice cream cone. We flew the ground pickup tests at Edwards and the water pickups at the Salton Sea near El Centro NAS. The land pickup used a goal post or mast station, and the water recovery station was a raft or small platform with a pole, as described earlier in the C-119 water pickups.

Approaches to both recovery stations were made at about 40 feet above the ground or water with a sharp pull up at contact. Again, as I described earlier, it was always fun to sit all the way back near the pole actuators as the "safety man" and watch the ground or water whip by, then ride the climb out. The safety man warned the pilots when it looked like the poles were getting too close to the ground or the water, and he also reported on how the recovery went after contact. Sitting back there hearing all of the noise and feeling the impact of the contact translated through the dolly did make me think about the tremendous forces that were being incurred by the winch and recovery rig as we hit these stations at 125 knots and bit into 800 pounds or more of metal. Reeling in these packages wasn't a big problem, although the ice cream cone sometimes started wobbling around as it neared the mouth of the fairlead. We then had to dampen the oscillations by reeling it in with coordinated winch and dolly movement, along with a fair amount of manual input from the loadmasters back on the ramp.

It was during a water recovery of one of these capsules at the Salton Sea that a heavy capsule submarined during the winch payout as we went into our climb. Capt Curtin was flying the aircraft, which was fast approaching a stall because of the drag of the capsule in the water. He hollered something like, "Cut it! Cut it!" to the winch operator, Willie Culpepper, who cut the cable, and as Don says, "Saved the day." Actually, I believe it also took some pretty good flying and instantaneous reflexes to keep us from becoming a C-130 submarine. We experienced a lot of incidents like that where unexpected things happened fast and we had no idea how close we had come to disaster.

One of the really novel recovery systems we tested at the Salton Sea consisted of a remote controlled catamaran that we dropped near one of our small capsules in order to recover it. The fiberglass boat was around 12 or 15 feet long and maybe 6 feet wide. It had a pole-type recovery station that could be raised on command and five or six cannons arranged in an array on the bow of the boat that fired 20 millimeter projectiles that were somehow attached to a net.

In the recovery sequence, we'd first drop the boat out of the aircraft by parachute so it landed near the capsule. After the boat landed in the water, the pilot began circling the boat. Using a remote control system, an engineer up in the cockpit of our aircraft would cut the drop parachute free of the boat, then start steering the boat toward the capsule in the water. When he got within "cannon range" of the capsule, he'd fire the net over the capsule, then begin backing up the catamaran to cinch the net closed around the capsule. Next he'd raise the pole station and we'd come along, snatch the capsule out of the water and reel it into the aircraft. After getting the capsule on board, we'd try to recover the boat.

The pilots flew the boat recovery as they'd fly any water pickup. I can't remember exactly what happened to the pole station on the boat during the recovery of the catamaran, but the boat usually came out of the water and trailed behind the aircraft nicely until it got to within - I'm guessing—maybe 100 feet or so of the ramp. The boat would then start swinging in wide circles and jerking about. It just didn't want to fly right. We tried various maneuvers running the rig in and out along with the loadmasters on the ramp trying to control it by hand, which at times almost got us beaten up by the winch line. I don't ever remember being able to bring the boat on board and we redropped it.

One of the more serious incidents that occurred during these water recoveries was when our T-28 chase plane crashed into the Salton Sea. Throughout our test program, aerial photographers in chase planes documented our recovery operations. T-28s, T-33s or T-37 "Tweety Birds" were the usual chase aircraft. At times these pilots would fly these planes right up to the rear of our C-130s to allow their cameramen to get good close-ups. We always had a lot of fun sitting on the edge of the ramp gesticulating, if you will, to the enlisted photographers who we knew personally. I kid you not; one pilot closed on us in a Tweety

X-15 research aircraft, top, in flight; bottom, with landing skids down./Photos courtesy of the NASA website

Bird until he was nearly right under our tail. How he got through all of the turbulence we generated beats me. You could hear the T-37's telltale screaming whine over the slipstream and our very loud jet exhaust and prop noise. I bet if we had taken a running jump off of the ramp we'd have probably landed on the Tweety's nose.

I was the flying safety man at the back of the aircraft when the T-28 went in. We were at about 40 feet of altitude and he was flying lower than we were. I was concentrating on the end of the poles to see how far they were above the water. I was watching the poles when out of the corner of my eye I saw what looked like the T-28's prop ticking the water. The plane climbed a bit, then nosed down into the water and flipped over. Capt Jarvis Adams was flying and I shouted into the mike something like, "Holy……. ! The chase plane just crashed!" Jarv said, "What?" I repeated that the chase plane had crashed into the water. Jarv immediately did a 180 and backtracked toward where the T-28 should be. We flew over the plane and saw that the pilot and cameraman were out of the aircraft and floating in the water, apparently OK. The pilot probably got fixated on us and also lost altitude reference because of the calm water over which he was flying.

Another one of our unusual recovery tests involved trying to recover the ventral fin of the X-15 research plane. The X-15 landing gear consisted of a nose wheel and two landing skids on either side of the ventral fin at the rear of the aircraft. The lower part of the ventral fin had to be jettisoned before the landing skids could be used. It was normally dropped and parachuted to Rogers Dry Lake just before the aircraft landed there. The top photo on the previous page shows the X-15 in flying configuration, and the lower photo shows the rear skids extended for landing after the lower part of the fin had been jettisoned. The part of the fin that is jettisoned can be clearly seen as the flat black portion of the fin in the top photo.

We flew recovery tests of the 10-foot long wedge-shaped fin in November of 1963. For these tests the fin was dropped from a C-130, not the X-15. The fin, although an aerodynamic body, behaved much like the catamaran in the water pickups. It would not fly in trail properly, maybe because it was relatively light, and we had a problem in trying to get it into the aircraft when it was reeled into close proximity of the rear of the C-130. Engineers tried to dampen the oscillations by adding a drag chute that pulled in the opposite of our direction of flight. I can't remember for sure, but I don't think this helped either. The fin still spun and flew erratically when it got into the wash and turbulence near the rear of the C-130. What I do remember for sure is that we were flying the tests on the morning of November 22, 1963, the day President Kennedy was shot. I happened to go up to the cockpit while the package was still in trail, and I saw Capt William "Vip" Vipraio suddenly take off his headsets and say that the president had been shot. We dropped the fin, left the drop test range and landed at Edwards.

I flew on those tests in the C-119 to try and retrieve objects on the ground, or in the water, by deploying a long line and flying in a tight turning circular flight pattern. We used significantly more rope than we did for standard recoveries. As I remember, the purpose of the system was to reduce G forces in recovering objects. The idea was to pay out a lot of line and then start flying in a turn while maintaining altitude. Initially the line trailed in a big circle, but then, because of gravity, the recovery end of the line began to spiral downward and turn in ever smaller circles. Eventually, the recovery line would reach the ground, tracking in a very small circle, at a relatively low rate of speed until it contacted the package.

So picture an aircraft trailing a line that has assumed a spiral shape with the small point of the spiral on the ground turning in very small circles. At contact, the pilot would then either straighten out or start a climb while still maintaining the turn, and the package was supposed to spiral off the ground and eventually, in kind of reverse fashion, un-spiral and eventually fly in trail of the aircraft like a normal package. I cannot remember how long it took us to fly each of these test recoveries, but it was a good bit of time, and it took some good and tiring flying to continually maintain altitude and rate of turn. I also can't recall if we ever successfully recovered one of those packages from the ground. I know I spent a fair amount of time looking at the horizon at a 30-degree angle or so from the back of the aircraft. I am still thinking that we might have done some of the tests at one of the dry lakes at Edwards. For some reason, I think I remember seeing the line leaving a trail of dust on the ground and it reminded me of the roadrunner in the cartoon. We only flew a few of those tests because they didn't work out.

As part of our job we flew on other types aircraft besides the C-119s and the C-130s. I logged time in DeHavilland L-20 Beavers and Cessna U3A Blue Canoes while looking for packages that went down on the drop range. The drop range was across Rogers Dry Lake from the main base at Edwards. Most of the time we drove from the base to the drop range in our crew pickup truck to recover packages. It was a long drive on that large base. Several times we flew out to the range in a Piasecki H-21 helicopter (nicknamed the "Flying Banana") to recover packages and fly them back to the base. It was a lot quicker and a lot more fun than driving a pickup truck.

We frequently flew light package drops from Gooney Birds (C-47s) and C-54s that were either assigned to Edwards or had flown up from Los Angeles. In my mind, I pictured these airplanes flying the air drops at Normandy or hauling cargo into Berlin during the airlift. The crew chiefs would remove the doors for drop and we'd simply toss our capsules and JATO bottles out the side of the aircraft. The packages would easily clear the tail of the aircraft.

```
                                    SECURITY CLASSIFICA
       JOINT MESSAGEFORM             UNCLASSIFIED
              SPACE BELOW RESERVED FOR COMMUNICATION CENTER
```

PRECEDENCE	TYPE MSG (Check)	ACCOUNTING SYMBOL	ORIG. OR REFERS TO	CLASSIFICATION OF REFERENCE
ACTION ROUTINE	BOOK MULTI SINGLE	AF		
INFO ROUTINE				

FROM:
6594 AEROSPACETESTWING OL NO 1 EDWARDS AFB CALIF

TO:
6594 AEROSPACETESTWING SUNNYVALE CALIF

INFO   6594 RECOV CONTROL GP HICKAM AFB HAWAII

SSD LOS ANGELES CALIF

UNCLAS FTZTW 24-9-62.

FOR TWGE, 6594 ATW; TRR, 6594 RECGP; SSOR, SSD. 24 SYSTEMS

DROPPED, 17 RECOVERED. OF THE SEVEN LOSSES (1) CHUTE FAILED

TO INFLATE (NOT RECOVERABLE). (2) CHUTE CAUGHT ON TAIL-WHEEL

OF DROP ACFT (C-47). (3) GOOD ENGAGEMENT, CHUTE LATER PARTIALLY

INFLATED AND PAYED OUT ALL THE LINE. (4) TEAR-THROUGH (BACKUP

COULD HAVE WORKED). (5) DE-RIGGED (BACKUP COULD HAVE WORKED).

(6) NO CONTACT. (7) NO CONTACT. ALL CHUTES WERE IRVING CHUTES.

POLE SPREAD TESTS - 15 DEGREES AND 10 DEGREES. POLES WERE

TEFLON WRAPPED.

DATE 251710Z
MONTH 24
SEP 1962

SYMBOL: FTZTW

TYPED NAME AND TITLE: DONALD R. CURTIN, CAPT, USAF
PHONE: 26831   PAGE NR. 1   NR. OF PAGES 1

SECURITY CLASSIFICATION: UNCLASSIFIED

SIGNATURE: Edward H. Mosher
TYPED NAME AND TITLE: EDWARD H. MOSHER, Captain, USAF, Officer in Charge

DD FORM 173, 1 MAY 55   REPLACES DD FORM 173, 1 OCT 49, WHICH WILL BE USED UNTIL EXHAUSTED

Recovery disposition report provided by Charles Dorigan.

However, on one Gooney Bird flight with Jim Stewart and me, things went differently. We tossed out the package, reported it to the pilot, but the recovery aircraft didn't see a parachute. After some time, when no one reported a package sighting, the Gooney pilot called back and asked us to look at the tail because he was feeling a buffeting in the controls. Sure enough, there was the parachute of the 150-pound JATO bottle we dropped hung up on the tail wheel of the airplane. A chase plane photographed the package trailing from the tail wheel, and it is now on video tape for all to see. The pilot landed the C-47 on the dry lake without incident or damage to the tail wheel. We removed the package, put it on the Gooney, and taxied all the way back to base ops. We had made the one and only C-47 Discoverer self recovery! The pilots at OL-1 prepared periodic recovery disposition reports that listed the number of successful recoveries, recoveries missed, and why they were missed. On my discharge from the Air Force, Jarv Adams gave me the one listing the tail wheel catch as a souvenir. I include it here (see photo on previous page) as an example of how we reported capsule catches and misses during our recovery operations.

As with the 6593d patch, Dorigan was asked to create the nose art for his C-130A, tail #131 at Edwards AFB. He did so, and provided a picture below showing his work. You can see the "falling star" in the nose art that the aircrew would "catch" during their missions.

When we first arrived at Edwards, Capt Ed Mosher thought we should be called "Star Chasers Extraordinaire." He asked me to draw a star with a comet's tail and paint it on C-130 #131. SSgt Lawrence

The falling star nose art on the C-119 of OL-1 at Edwards AFB about 1959. Members of OL-1: Left to right: (kneeling) Capt Ed Mosher, Capt Donald Curtin, Capt Jack Parker, Capt Floyd Barrow, TSgt Frank Kenyon, and SSgt Doug Stinnett (standing) SSgt William Culpepper, TSgt Billie Hendon, A2C Jim Stewart, A1C Walter Johnson, A2C Donald Hackworth, A2C Charles Dorigan, SSgt John Kosmatka, and SSgt Ken Klein./Photo credit: USAF

Bradley and I cut a stencil out of some cardboard and spray painted a blue shooting star onto the side of the aircraft. That symbol was also painted on our C-119 and our first C-130B, #526. I've often wondered if that was the origin of this "Catch a Falling Star" business, the unit motto of the 6594th Recovery Control Group. It probably wasn't—there was a song Perry Como sang by that name.

Capt Mosher also wanted everyone in our detachment to have a uniform nametag for our flight fatigues. The ones we chose were the leather nametags with silver letters and wings. These tags slid into soft, transparent plastic holders that could stand up to cleaning and such. Seen in the photo on the following page, everyone in that photo except one was wearing one of those nametags. You can also see the shooting star symbol on the C-119 behind us. Our ball caps were robin's egg blue with dark blue numbers, and we painted our P-4 helmets the same light blue with dark blue trim. The 6593d's ball caps were dark blue with white numbers. Later, Ed asked me to design a recovery patch for our detachment and I drew the C-130 catching a parachute.

Dorigan left the Air Force in 1964 and returned to Hawaii and attended the University of Hawaii. He actually used Corona footage during his defense-related career before he retired. Dorigan looks back on his years in aerial recovery quite fondly.

> *"The years at Hickam and Edwards were some the best years of my life. I worked with great enlisted men and officers, and at Edwards I witnessed whatever was new and exciting in aviation."*

The years at Hickam and Edwards were some the best years of my life. I worked with great enlisted men and officers, and at Edwards I witnessed whatever was new and exciting in aviation. I was career-oriented and hadn't planned to get out, and in fact was reluctant to do so, but events in my life caused me to change my mind and I started to think about pursuing a full time college education. I applied to and was accepted by the University of Hawaii. I was discharged on July 13, 1964, and left for Hawaii the following week. Even though I was out the Air Force, I wasn't quite out of OL-1. Several OL-1 people rotated back to the 6593d in Hawaii and we kept up our close ties. They provided me with a great deal of moral support as I started and made my way through college. My major was in physical geography and I took courses in aerial photo interpretation, cartography, earth sciences, and climatology.

My education and interest in looking at the earth from above led me to working in remote sensing where I interpret photography, satellite images, and thermal and radar imagery for environmental, natural resources, intelligence, and defense-related applications. We use Corona information from time to time where I now work at Earth Satellite Corporation in Rockville, Maryland. In fact, the director of our Department of Defense and intelligence programs at EarthSat, Jim Fry, performed photogrammetric measurements on Corona photos. Corona imagery provides us at EarthSat and other scientists with valuable baseline data that we can use to monitor the changes that have occurred in the earth's landscape and urban environments since the imagery was first acquired. Nowadays the Corona films are usually digitized for analysis, but I keep a canister of old Corona film at my desk as a reminder of what we did so long ago and how far we have come since then.

In 1995, much of the Corona data was declassified and a declassification ceremony was held at CIA headquarters in Langley, Virginia. I had the pleasure of taking Jarvis Adams to the ceremony at the CIA and to the reception held that night at the Air and Space Museum.

Veterans of the 6593d Test Squadron (Special), Charles Dorigan (left) and Daniel Hill at their 2002 unit reunion./Photo credit: Daniel Hill

The original members of the 6593d have held reunions every two years since 1990, and I hosted the reunion in 2002 in Washington. We have a pretty decent network of people in communication with each other. A couple of years ago, our group was registered as an official non-profit historical organization in the state of New Hampshire, the 6593d Test Squadron (Special) Historical Society.

Chapter 6

SMSgt Richard C. Bell

SMSgt "Casey" Richard C. Bell at Patrick AFB in 1964/Photo credit: Richard Bell

"Space capsule recovery was very routine for the loadmasters."

SMSgt Richard Bell (1931-) was interviewed using e-mail between May and July 2003. As an airman first class, Bell was an experienced loadmaster when he was first assigned to the 6593d Test Squadron (Special) in August 1964. He started as a JC-130 recovery loadmaster and progressed into a winch operator. He was involved in the recoveries of both Corona capsules and Ashcan payloads. Bell was promoted to staff sergeant (E-5) and tech sergeant (E-6) while he was assigned to the 6593d. Bell was a career airman and was transferred from the 6593d in 1968.

Bell joined the 6593d in 1964. He notes that the assignment was very intriguing. Although he didn't recall much being revealed in briefings about Discoverer, Bell knew the aerial recovery mission was a top priority.

I heard about the 6593d aerial recovery operations and it sounded interesting. A recovery crew from the unit at Edwards Air Force Base (AFB) gave a demonstration recovery at Patrick AFB, and I thought I'd like to try it, so I volunteered. Being in Hawaii was no small part either.

I don't recall any specific briefing about the purpose or mission of the Corona capsules we were recovering. We didn't know much, but I was quite sure they had a photographic capability.

We were told the mission had the top Air Force, Department of Defense (DoD), and National Aeronautics and Space Administration (NASA) security. When we were on Temporary Duty (TDY) the back of our flight orders were stamped with a statement to get us any support we required. I can't recall the exact terminology, but on the back of our flight orders stated our mission was "Air Force Priority XXX, DoD Priority XXX, NASA Priority XXX." We were told the "XXX" was the top priority for each department.

Prior to my arrival, there was lots of publicity. The Discoverer program was even reported on local radio stations when a recovery had been made. They took a photo of the crew and capsule. Someone gave a talk about it at a local church, and it was in the newspaper.

I'd guess the timeframe as 1965 when a popular men's magazine had quite a story on the function of the capsules. Perhaps the magazine's name was *TRUE*, but it has been out of print for many years. We were all called into the briefing room and told not to discuss the story.

Initial training started with learning about the recovery equipment. Bell describes how they made loops and added the hooks. He also notes the color codes used for the equipment. Even in training, problems could occur. Bell shares one particular training day.

The first part of loadmaster training was learning to splice and make the "loops,"—the trapeze-like line that was used to recover the Corona parachutes from a JC-130. It could take a couple of hours to make a loop. They were very particular with the splices. I think there were a total of sixteen splices. We used half-inch nylon rope.

The "rig" was the rope/hook assembly attached to the recovery winch line and the poles. The parachute was captured by contact with the rig. The nylon rope used for a capsule recovery came in 600-foot spools. We used 500 feet for the winch line, and made the loops from the balance. We had a "schematic" on the hanger floor to cut and standardize the various lengths required. Starting at the winch line end, we'd splice an eye in two lengths to connect the winch line with a clevis. Next came the "in-line" hook—a lateral line with a hook in the center, connected to the first two lines, like a trapeze. We cut two more lengths with a byte spliced in, and a second lateral line with an in-line hook. The bytes were connected to additional pole hooks. The splices were what held the whole thing together. Oh yes, the lateral lines had a cotton cover over them, to reduce the friction from nylon-on-nylon contact. I suppose much of the emphasis on making a good loop was the fact it was only used for one mission recovery.

The recovery lines/loops were quite sturdy. I'm guessing we didn't have more than one or two break during my four years, and they were all training loops that probably just wore out. I don't believe we had any limit on how long a training loop was used. Some looked quite shabby before they were trashed. We'd just cut the hooks out and use them in making another loop. There really wasn't much in the way of strain during the recovery. The winch drum turned four free revolutions before the brakes were applied, and the 150 feet of payed out winch line had a great deal of elasticity in it. I once heard an All American tech representative say there was less shock load on a capsule than on a personnel chute deploying. I don't really know if it was true, but he said it.

For an actual satellite capsule recovery mission, the winch line, loop, hooks, and poles were all brand new. As each item was assembled for a mission, it was color-coded green. The pole ends, loop ends, and the hook-retainer assembly were all marked green. After a mission, all were relegated to training equipment and repainted red. The winch line was spliced to the drum and used for twenty practice recoveries, then reversed for another twenty. The drum was the cylinder the winch line was wound around, like the spool on a fishing reel. The splices could be viewed as the weakest link, I suppose.

During our flight training missions, we used 200-pound "bombs" filled with metal shavings to simulate the capsule. There were four "riggers" (or loadmasters), two forward of the dolly, and two aft of the dolly near the JC-130 recovery ramp. They'd start you in the forward position, you'd mostly observe while making a couple of recoveries from there. Then they'd move you to the aft position. The left-rear rigger was the primary; he was the one controlling the functions from back there, in coordination with the winch operator. I think most riggers got their check ride after ten to fifteen training flights. You were not assigned to fly an operational recovery mission until after a check ride to become qualified. Once a rigger was qualified, he was usually put on the next recovery mission, but worked a forward position for several missions. There were no hard, fast rules on it.

JC-130 during an aerial recovery in 1962./Photo credit: USAF

The winch operator was in charge of the riggers. There wasn't a great deal of leadership required. You'd just tell the riggers which position you wanted them to man. However, on my crew, I didn't even have to do that. They just rotated positions automatically. Rank was the primary requirement to upgrade to a winch operator—all were at least staff sergeants. As soon as I made staff sergeant, they upgraded me to a winch operator. It only took a couple of flights to get qualified as a winch operator. You already had a working knowledge of what was going on.

I was in the 6593d Test Squadron from 1964 to 1968 when there was a shortage of loadmasters due to the Vietnam War. We got most of ours straight out of the basic loadmaster school. The loadmaster requirement goes way back to a project flown with C-119Js in the mid-1950s (Genetrix project). The 6593d didn't have any women in the recovery aircrews. I don't think the first women came into the loadmaster field until the late 1970s.

We used 200-pound bombs, but had them weighted to 180 pounds. We had bomb shackles on the dolly, and for training we dropped the bombs to ourselves. When we went to the 900-pound and 1,800-pound training capsules, we had to have a drop plane and a recovery plane. It took considerably longer to deploy the rig for heavy weights. The loop was made from either three-quarter-inch or 1-inch rope, I forget which, and instead of the half-inch nylon winch line, we used 1,000-foot stainless steel cable. After the rig was deployed, we had to put down a set of "dog houses" (three-sided "boxes" of plywood covered with aluminum that attached to the rails) and created a tunnel over the cable. The recovery equipment

JC-130 in 1961 with its dolly in the rear position on the ramp and the recovery poles without a loop. This is most likely at Edwards AFB photographing the aircraft and recovery equipment, not a recovery.
/Photo credit: USAF

was capable of recovering at least 3,000 pounds. They did that at Edwards AFB, but never tried anything heavier. As the 3,000 pounds came onto the ramp, the airplane began to lose stability from the tail-heavy condition. I personally never was involved in any capsules above 1,800 pounds.

I can't remember the first time I participated in a training parachute recovery. Space capsule recovery was very routine for the loadmasters. I can recall the first (and only) failure I was on. I don't recall the entire crew, but the pilot, winch operator, and left rigger were all from Standardization. The capture (from the view on the C-130 ramp) went like this. When you heard the ten-second warning, then you grabbed hold of the side of the aircraft to prepare for some last minute rudder-kicking. As the chute went under the ramp you called, "Contact" on the intercom, and approximately one second later, "In trail." Well, on this mission it went more like, "Contact…tear-through." The response from the pilot was, "Oh, (curse word)!" That's all that was said over the intercom (excluding checklists, of course) for the remainder of the flight. Another aircraft was close behind us. They made the recovery within a minute or so of our parachute tear-through.

A "streamered" parachute was a failure that prevented the chute from opening, possibly due to improper packing or perhaps getting wet. They just failed sometimes, even with personnel chutes. Perhaps the pole hit the outside edge of the chute, sort of a glancing blow with the pole, one hook catching a lateral band of the chute and closing it up like an old tobacco sack. The chutes were called "ring slot" chutes. That is, it was not a solid canopy, more like a series of bands around the circumference of the canopy. These bands of material were sewn to reinforced lateral bands, and I believe every other suspension line (from the canopy down to the attaching point) was reinforced. Perhaps "reinforced" is misleading, "heavier than normal" may be a better description, but they were known as "reinforced."

We had a higher-than-normal rate of "steamers" among our practice chutes at one point in time. A parachute engineer came out to check them and discovered they were not manufactured to specification. I'm not sure of the exact spacing between bands, but for the sake of clarification—let's say 6 inches. He found some with 6 ¼, 6 ⅜, etc. It didn't sound reasonable to me that this small difference could cause a problem. He explained it by using a paper bag as an example. Punch a couple holes in the bag with a pencil, and the flow of water would still fill the bag. Punch a couple more and the loss of water through the holes would prevent the bag from ever filling. In the case of a real capsule returning from space, a malfunction that caused the chute to deploy at too low an altitude and with too great a descending speed could blow out some panels. I'm not aware of that ever happening.

Edwards AFB had two or three recovery-configured JC-130s involved mostly in test operations. Also, at this time the Air Force had a parachute test unit at El Centro Naval Air Station. For heavy drops/recoveries, when a second plane was required, El Centro crews did a lot of dropping to the Edwards crews. During this one particular test of a large cargo chute (probably a 64-foot chute, I have no idea what they were looking for in the test), it was a drop only, and the Edwards crew was just photographing the sequence. The El Centro aircraft commander asked if he could make a "dry run" over the chute during its descent, just to see what it was like to line up on a chute. Since no recovery was to be attempted, the Edwards aircraft commander said, "Sure, why not?" Remember, this was being photographed and I saw the film. The El Centro aircraft commander flew his C-130 right into the chute. The parachute exploded like a kid's balloon. The canopy streamered around the nose and covered it like a cloth draped over a fence post. The Edwards aircraft commander had to guide him around a bit until the chute all ripped and blew away.

As Bell describes it, he was not permanently assigned to one pilot, or one C-130. Bell explains one mission shortly after his arrival.

We were assigned to crews. The winch operator was the reporting official for the four riggers in their Airman Performance Reports, so we tried to always keep them together. We flew with various pilots. The pilots in upgrade training had no assigned crews, so we'd have to cover their training flights. Maj William Vipraio was my aircraft commander until he rotated back to the mainland. I also flew missions with Capt Curtin, Maj Bayer, Lt Col "Tex" Owen Pratt (he was the commanding officer of the 6593d), and perhaps Capt Douglas Sliger.

We seldom flew in the same JC-130 for two recoveries in a row. They were all the same to us. The JC-130 was the only airplane I flew recovery missions on, but from my viewpoint it was well suited.

Shortly after I arrived, a mission was running over a weekend. Since I wasn't qualified yet, I was the duty driver. The emergency forces had to show and cock their airplanes, then they had some time to kill while they waited for their release. There was a poker game in the recovery section—in the corner of the hanger, next door to the hanger that the group was in. Someone came into the hanger and shouted, "Emergency forces released." Everyone left except the players. In just a manner of minutes as the poker hand was still going, we got another shout, "Scramble emergency forces!" Pandemonium ensued. One of the aircraft commanders was gone, but Maj Bayer was there to check his mail or something, so they grabbed him. They grabbed me and this gave us three riggers per crew. Well, we blasted off, flew out a ways, and then got a release.

Bell explains the mission process and the schedule each crew member had to follow. He continues by describing how the recovery equipment was prepared for live missions.

This becomes involved, but I'll try to make some sense of it. The duration of the capsule in orbit might vary, but let's assume this was a five-day mission. We had our crews scheduled the day prior to launch. Now the schedule was set for day one with the assumption the capsule would be returned early. We were given Estimated Time of Parachute Deployment (ETPD), and everything was backed up from then. We wanted one hour "On Station" circling the expected recovery area prior to ETPD, add en route flying time, preflight time, and arrival at a briefing time. For day one, crews one and two were "emergency forces," and crews three, four, and five were considered "normal forces."

At a point in time, the folks in Sunnyvale decided that the capsule was operating normally and it was go for day two. At that point, it was too late to operate with the schedule as described, so the normal forces were released. The emergency force was kept until the last minute, so they were able to be On Station without the one hour prior. I'm sure this is confusing. It took awhile for all of us to really get in the flow—no other operation is scheduled in a manner such as this.

For instance, assuming a 7:00 a.m. briefing, the crews figured on leaving home at 6:00 a.m. Early on, the winch operator attended the briefing, but it really was a waste of time for us. It dealt with the route, weather, and etc, so at one point we no longer attended. Briefing time for the recovery crew was when we began getting the equipment and doing the preflight.

If conditions warranted, the normal forces were released prior to 6:00 a.m., the aircraft commander was notified, and he released the copilot, navigator, flight engineer, and the winch operator. The winch operator passed the release to the four riggers. Often times everyone showed at the squadron at 4:00 a.m. to find out we were released! The same routine was followed until day five, when we were all told that day was primary. I'll lay out a timeline to make it clearer: 7:00 a.m. briefing; recovery aircraft take off at 9:00 a.m.; two hours of flying time; On Station at 11:00; and ETPD at 12:00 local time.

Emergency forces would definitely show and the airplane would be "cocked"—that is, the preflight was done, up to the point where "start engines" was the next step. Then the crew could leave the aircraft and seal the door, and stand by until their release. They would be able to "scramble" (continue the checklist) and take off as late as 10:00. Normal forces could have been released at 6:00. Of course, we (the crews) were not involved in all of this planning, as far as when it was primary. Everyone could possibly be released at 6:00 a.m. You just assumed you were flying that day until the release came.

We had five aircraft on a Corona capsule recovery mission. The airplanes were lined up numerically from north to south on the trace, which the capsule was expected to return on. Both pilots and winch operators normally flew their first mission in the number five slot and worked their way up.

For a mission recovery, first, equipment was drawn from unit supply, to include a loop, winch line, re-drop chute, ditty bag (containing two pole hooks, masking and electrical tape, clevises, safety wire, and etc), and a capsule canister. All five crews loaded this on a truck, loaded mission poles on a trailer, and then went to the flight line. The practice poles were removed, and the mission equipment was loaded on each aircraft in turn.

We loaded the poles in the hydraulic actuators, the part of the dolly that held the poles and raised/lowered the poles. The dolly had its own electrical/hydraulic system and it could be operated without the JC-130's engines running. The dolly was the large, yellow piece of capsule recovery equipment that you see in some photos. It traversed forward and aft by an electric/hydraulic system that was self-contained

Loadmasters standing beside a JC-130 recovery dolly./Photo credit: Richard Bell

in the dolly. The winch hydraulic motor was powered by the aircraft's hydraulic system, so we did not remove the training winch line and install the mission line until after takeoff. If we had a short en route flying time, we would sometimes begin the removal during taxi. If the winch operated during taxi it overheated the hydraulic system, so we could only operate the pumps in "low," taking twice as long. The hydraulic system for the winch was an add-on to the aircraft system. It had a cooling system (think of a radiator on the bottom of the wing), but was designed to operate in flight. After removal/installation of the mission winch line, we installed the pole hooks, attached the loop, and we were ready to go.

Among the equipment, I mentioned a "re-drop" chute. In the event of a malfunction with the winch after getting the capsule in trail, and being unable to winch it aboard, we had a re-drop procedure. We always had a re-drop chute and "pig tail" (a 24-foot piece of the nylon rope used for the rig/winch line with an eye spliced in each end). We could attach the pig tail to the winch line at the forward end of the dolly trough with a couple of cable clamps. We attached the re-drop chute to the eye of the pig tail, secured the static line, cut the winch line, and re-dropped the capsule. As far as I know, it was never done on an actual Corona mission. We did it with an Ashcan parachute over Alaska one time.

During the winching in, with the Ashcan parachute perhaps 50 to 60 feet in trail, the winch hydraulic line blew out an O-ring seal so we couldn't use it. We connected a re-drop chute (a regular training chute) to the winch line with cable clamps while a ways out of Eielson AFB. We just towed the parachute until we were over the field at Eielson, flew a low approach, cut the winch line and dropped the Ashcan alongside the runway. It worked just like it was supposed to! This was the only time the procedure was used, as far as I know. A ground-controlled interception station along the way called us on the radio and asked, "Um… Ozzie 56, is that a normal configuration for your aircraft?"

Every crew member had specific job functions during a mission. Bell concentrates his descriptions to loadmasters and winch operators. He also recalls the play-by-play of a live recovery mission.

The winch's brake and free turns were pre-set prior to the beginning of the sequence, and enough slack was put in the winch line to allow the rig to be deployed. "Brake and free turns" describes the mechanical sequence of the brake application during the recovery sequence. The recovery winch used the same brakes used in C-54 aircraft. There were two brake settings used for Corona missions, an "initial" and a "final" brake setting. The initial was approximately 1,800 pounds of brake force, and the final was approximately 3,000 pounds of brake force. "Free turns" were the number of turns the winch drum rotated prior to the brake application. This method decreased the shock loads encountered. My description of what was actually happening would be to compare it to the parking brake on your car. If you set the parking brake, the car will not roll. If you set the parking brake and added four "free turns," and your car was parked on a slope, the car would roll until the wheels revolved four times prior to the brakes stopping and holding the car. After the winch line had payed out and stopped, the final brake settings were dialed in prior to winching the capsule in.

The first step in a recovery was for the aft left loadmaster to open the ramp and door. The poles were horizontal and about two feet off the floor. As soon as the ramp and door were open, the aft left loadmaster checked with the aft right loadmaster, and then gave the winch operator a hand signal to "milk" (lower) the poles down. That is, the winch operator turned the dolly hydraulic pump on/off, putting just a bit of pressure in the system and lowered the pole tips just above the ramp, making it easier to deploy the rig.

When cleared by the pilot to deploy the rig, the aft left loadmaster hand-signaled the two forward loadmasters to manually push the poles from their stowed position at the front of the dolly to the rear. The two on the ramp hand-fed the loop out as the poles and dolly moved aft. I mentioned the slack in the winch line. It was tied off to a ring on the winch. As soon as the poles had been pushed to the rear position, the forward right loadmaster freed the winch line and allowed it to play out the slack as the dolly moved aft. As the dolly was far enough aft, the aft left loadmaster hand-signaled it was OK for the winch operator to begin moving the poles to their down position. As you see, most all of our deploying sequence was done with hand signals. We knew what the next step would be, and it was just easier to signal rather than use the intercom. The aft left loadmaster announced, "Good rig." The sequence probably took from three to four minutes.

We tried to be rigged and ready to meet the capsule at 15,000 feet. First, there would be a visual pass of the descending parachute. Usually, the pilot visually checked from the swivel at the top of the capsule harness up to and including the chute. The engineer checked the harness and capsule. Both looked for any indication of anything broken or missing that could hinder a good recovery.

The pilot announced on the intercom that the chute was passing the left wing tip. The aft left loadmaster announced over the intercom when he had the chute in his view. He made calls to assist the pilot maintain his desired rate of descent. Something like, "It's below the horizon and rising—below the horizon and rising—coming up on the horizon—it's on the horizon and steady." Ideally, the chute's skirt appeared to be directly on the horizon. The copilot called out the time since the wing passage, normally after about thirty seconds, and the pilot announced when he was beginning his turn. The aft left loadmaster made his final call, "On the horizon and out of sight."

When the aircraft lined up with the parachute, the pilot announced, "Inbound hot." The winch operator responded, "Winch operator check complete." The pilot gave us the "ten seconds warning" and "under the nose" announcements. This was a signal for the aft loadmasters to hang on and prepare for some serious "rudder-kicking," which is sort of like a car skidding. You pretty much knew the pilot and his reputation, and that indicated exactly how tightly you held on. There were just a few "rudder-kickers" that you were always prepared for. They were about ten seconds out, saw that they were not aligned too well, and made corrections. The ten-second warning was primarily for the flight engineer to start the cameras and the winch operator to be alerted. It was a secondary feature as far as warning the aft riggers. The best pilot I flew with was Maj Vipraio. You didn't have to hang on at all when he was flying.

> *"The best pilot I flew with was Maj Vipraio. You didn't have to hang on at all when he was flying."*

Assuming a good catch, the aft left loadmaster announced, "Contact…in trail." The winch operator notified the copilot about the length of winch line payed out for the recovery report. The pilot would clear us to bring the parachute and capsule in. The aft left loadmaster kept the crew advised about whether the capsule was trailing smoothly, oscillating, or whatever. He announced as the chute began to come through the horse collar on the end of the dolly, counting on the intercom, "five, four, three, two, one, and stop" as the capsule came up to the collar. We'd raise the poles, raise the dolly trough (to keep the capsule off the floor), move the dolly to its forward position, secure the poles, and with an "OK" from the pilot, closed the ramp and door.

An aft JC-130 loadmaster standing between the dolly to his left and the extended recovery pole to his right. The red training payload dangling at the left has just been recovered and reeled up to the fairlead of the dolly.
/Photo credit: USAF

An aft loadmaster standing on a dolly as a parachute and capsule are being recovered by a JC-130.
/Photo credit: USAF

During the first few missions I flew, we untangled the loop from the chute, placed the capsule in the canister and locked it. We wanted our loop back, and occasionally you had to cut parts of the chute to get it free. Coincidentally, this made some scraps of chute and sometimes they might be kept as a souvenir. Then the people at the other end wanted the chute and loop intact and placed in the canister with the capsule.

The successful aircraft parked, and opened only the crew entrance door. The commanding officer normally rode out in the crew bus, or 1½-ton truck in our case. He'd always get on board, walk around shaking hands and giving us what seemed to be a canned speech on training, standardization, and etc. We were all taken in to a debriefing. A couple of administrative guys from the group headquarters office came out and got the capsule canister off and took it to an aircraft waiting to take it back to California. It was usually returned in a Military Air Transport Service (MATS) C-135, and those crews referred to it as a "pineapple mission" when returning a Corona capsule to the mainland.

There were no security forces involved. We were always told it was done like this in the name of security. But, if someone really wanted to know—if you saw five aircraft load up and take off in a short period of time, and when they returned, four of the planes were opened up and all their equipment was removed, but one plane stayed closed and a pickup went to the other side, removed a canister and took it away—well, I've never worked in intelligence, but I think I could figure this one out.

We wore a parachute harness (no chute) that was connected to a restraining strap attached to the aircraft. Perhaps in the 1967 timeframe, they replaced the strap with a retractable reel assembly (sort of like the retractable extension cord trouble light some mechanics use). Also, about the same time, the aft riggers had to wear a helmet.

There was no formal winch operator training. Upgrades were too infrequent, and really not required. We were given a briefing on the winch components and operation by a tech representative from All American Engineering. Then we were given a demonstration during a training mission by a winch operator instructor, and then you began doing them under his supervision. A normal training mission had four recoveries: the aircraft commander made two, the pilots changed seats, and the copilot made the other two. Most winch operator checkouts could be done during one training mission.

The winch operator monitored the activities of the riggers, kept the pilot advised of anything not going right, ran the winch to its aft position when ready, deployed the rig, made the recovery, and brought the capsule back on board.

A log was kept on each aircraft to record the winch payout for each recovery and also the brake settings. The brake settings were calibrated by the Recovery Maintenance section. A large scale was attached to the floor of the cargo compartment, and we attached the winch line to the scale to determine the required settings. The number of recoveries per winch line was also kept in the log. For training, I believe twenty were made on a new line, then the line reversed for an additional twenty. For preflight you only needed to check the log and make a visual look-see of the equipment. The whole thing was quite simple.

The Recovery Maintenance Shop did the maintenance and repairs on the recovery winches. It was manned by tow reel operators. They were the career field that operated tow reels for target-towing. This was a defunct (as far as I know) mission where aircraft such as a B-26 Invader towed a target for other aircraft to shoot at.

The aft loadmasters lowering a recovery line from the C-130A over Edwards AFB in 1959./Photo credit: USAF

Bell describes the competition between the pilots. Recovery rates were very important. If a pilot did well, he qualified to be on the mission line up.

Among the aircrews, only the pilots were concerned about recovering the most capsules. I guess you'd call it an informal competition. They just liked to brag. But that's just a part of being a pilot, I suppose. It was almost fun and games, with a very serious overtone. The squadron had a recovery percentage rate. It meant nothing to me, so I never put any effort into keeping abreast of it. A pilot had to reach the "magic number" percentage during his training missions before he was put into the mission line up. He also had to maintain the recovery rate to stay in the lineup.

One of the pilots (a major) was never able to achieve it. It got to the point where he'd be physically sick on the mornings he was scheduled to fly a training mission. They finally removed him, made him the officer in charge of the Recovery Section (the loadmasters), and he only flew local training, touch-and-go flights, and etc but never any more recoveries.

I really don't have any idea how many Corona capsules my crew recovered. It never came up before. I didn't keep count on how many Corona recovery missions I was involved in, but a conservative count might be over twenty. Remember, it took five crews for a mission, but only one for a recovery. When I was promoted to tech sergeant in 1966, I was also given the job of running the Recovery Supply section. From then on, I wasn't assigned to a firm crew, but I flew when I needed to fill in.

Bell discusses the recovery poles on a JC-130. As sturdy as they were, Bell explains how quickly they could be damaged. He continues by providing information about the equipment during missions.

The poles had a baked-on finish, Teflon as I recall, and we had to wipe them down with a damping fluid prior to the rig being deployed. The fluid was in a paint can, and was very slick stuff. It had almost a greasy feeling to it. It was supposed to lesson any friction from the contact with the chute.

We had to jettison a lot of recovery poles into the ocean. If the pilot was off just a little and hit the chute directly with a pole, the pole would bend at about a 90-degree angle, and sometimes just snapped off. Either way, we didn't like to bring the poles back on board.

During training missions, we had two poles in the dolly and two spare poles were tied down in brackets on the right side. If a pole was bent badly, it was hard to handle on board and to secure in the stowage brackets. When they broke, the edges were razor sharp, so we didn't like handling them. As I recall, the pole weighed about 190 pounds, was segmented from approximately six inches at the large end, to four inches, three inches, and two inches at the retainer/hook end. It made them more difficult to handle with one end so much heavier than the other. Someone finally got the idea of repairing the poles to be used on practice missions, but only a small percentage was repairable. It was done by the base shops at Hickam. We also lost lots of the practice bombs due to streamered chutes (parachutes that did not open), tear-throughs (contacting a parachute but not recovering it), and de-rigs (separating the recovery loops from the poles) that we were not able to straighten out and re-deploy before we ran out of altitude. I've wondered what will happen when the ocean dries and some future generation wonders what the hell kind of war we fought with the poles and bombs filled with sand.

We had a camera mounted in the front windscreen between the pilot and copilot, as well as one on the bottom of the dolly pointed aft. The flight engineer had a control box, and turned them on at the ten-second warning. Each pilot was able to review some photographs of his previous day's catches. At some point, someone got the idea to paint the poles with black and white sections for better photo identification. They were painted at the factory. We got a few repaired poles back, and they were also painted black and white by the paint shop.

A lot of the Air Force operational theories were, "We did it that way yesterday, so it must be right, and that's how we'll do it tomorrow." Nowhere was that theory more practiced than in the 6593d. For example, although we flew daily training missions with two poles in the dolly and two spares in the rack, come mission day, off the spare poles came. We all understood we had to use the new mission poles, but we did not understand the removal of the practice poles. Reason given: a hazard! We fought that over and over—always lost.

Fade to the next mission. When one of the crews was initially trained, they had flown quite a few missions (all unsuccessful) by the time I was involved. During preparation for my first one, I discovered we now had to carry four poles: two in the dolly and two spare mission poles in the brackets. I was astonished. I asked the reason and was told there were only three aircraft used on these, and there was the possibility of making a ship data pickup. OK. But a few days later I gave it some more thought. If an aircraft made a successful recovery, it would have an object tied down to the floor, aft of the dolly, so it was physically impossible to deploy the rig for a ship data pick up, or for any other reason. I brought this to the attention of the operations officer and he said, "Good thinking." We'd go back to only the two poles in the dolly. I asked if we couldn't leave the practice poles on—we'd flown missions with them. "No" was the reply!

From my experience the word "change" wasn't in the program. We ran a test on about 250 test chutes that were just a bit different than the normal ones we used. We were told prior to trying the first one, it would not be used! I heard an operations officer say (when discussing a change of some sort), "I'm just glad they didn't fly the C-119 upside down when they made the first recovery."

Not all missions were a success. Bell describes three occurrences in which something went wrong. The resulting outcome was broken equipment, lost capsules, and injuries to the crew. Bell illustrates how and why his job was quite dangerous.

One capsule came down in the middle of a rain squall, and we never had a shot at it. My crew flew high cover orbits maintaining electronic surveillance, and I believe a second plane flew a low cover also. I'm not 100 percent sure on the second one. The remaining crews returned to Hickam, refueled, got new crews and replaced us as we got low on fuel sometime after dark. For fuel conservation, we shut down two engines and flew just above stall speed. You can't go anywhere on two engines, just do a very slow speed orbit. We had to restart the engines to return to Hickam. I really don't recall the fate of the capsule. The squadron had CH-3B helicopters with divers that were capable of water recoveries, but it seems this one was out of their range.

On another mission, the electronic direction finding operator picked up the capsule radio beacon in what seemed to be a normal reentry (coming north to south), but then he said, "Oh, it's continuing south." All five aircraft got the same signals, and all turned south, chasing it until we were told to discontinue the chase and return home. I don't recall the outcome of this one. On the day of the incident, the training capsule was dropped by another C-130. We were not ready in the rear, the poles were still horizontal and the brakes were "set" (like having your parking brake set). To get ready, we had to lower the poles, and set in the delay and initial brake setting. The pilot decided to make a dry run over the recovery chute before having us deploy the rig. As we crossed over, the recovery chute popped up, caught the rig, and pulled all the winch cable out. The cable was a seven-sixteenths-inch stainless steel cable attached to the drum with a nylon "weak link."

Anyway, the process didn't work like it was supposed to. When the payout began, it caused the winch to over-speed and the cable "grew" (expanded on the drum) from centrifugal force, apparently separating the end of the cable from the weak link. If you're a fisherman, think of a really bad backlash. As all this mess spun, it tore up the cable, which looked like a shredded brillo pad. It shot pieces of cable into some hydraulic lines (like the proverbial straw stuck in a telephone pole by a tornado), the sound roofing, and about everything else in the cargo compartment. The right forward rigger got a small piece of cable shot into his arm between the wrist and elbow, and I got a small piece in an eyelid. The flight surgeon pulled mine out, but said it was better to leave the small piece in the other rigger's arm bone. It also slashed the hydraulic reservoir on the dolly, which lost all its fluid and made the floor quite slick. We had to disconnect the dolly drive shaft and pull the dolly forward by hand. We also lost all intercom in the rear and we had a problem trying to keep the aircraft commander informed.

Perhaps this will clarify a bit. When using the nylon winch line, it was spliced to the drum. When there was a brake failure in the winch, we'd get a 500-foot payout, but the line remained attached. We could clamp it off to the dolly, drive the dolly forward to get enough slack to start rewinding the line, and then reel it in normally. The idea behind the weak link (it was approximately 50 feet long) was if there was a brake failure, all 1,000 feet of cable payed out, then the weak link just snapped. It really didn't work like the engineers conceived it!

There was a second incident. There was a little push-pull lever that had to be moved to the side of the drum that was used. The choices were either large-cable or small-nylon line. We put on a new cable one evening prior to going to Edwards. I moved the lever to "large," but it didn't engage. It's like engaging two gears that you're unable to see, and if the teeth don't mesh, the lever stops, and it's not engaged properly. Instead of being tooth-to-slot, it was tooth-to-tooth. Anyway, we tried to make a recovery and got a 1,000-foot payout. It did very minor damage, so the weak link did most of what it was supposed to in this instance. When we hit the chute, we had zero brakes! After that, they painted scribe marks on so you could be sure the gears were engaged properly.

Bell was involved with recovery projects other than Discoverer. He shares his experience with the Ashcan program as well as recovering telemetry data capsules and NASA's BioSatellite project.

The capsules sent out information on telemetry, primarily to ground stations. However, there were areas of ocean that were too far from the ground stations. Two or three ships were manned by civilians and equipped to receive the telemetry at sea. If the telemetry was needed in such a hurry that they were unable to wait for the ship to make port, they could load the telemetry data tapes in a buoy-shaped container about 4 feet high and 2 feet in diameter. They connected it to a balloon filled with helium, floated it about 100 to 150 feet above the ship, and our JC-130 crews could snag the telemetry data tape container and return it. My crew practiced this one day, and made three or four pickups. Prior to the final run, the captain of the ship asked how many crewmembers we had and was given an answer. He said to be sure to open the final capsule. It had a case of cold Pepsi rather than the concrete blocks in the others. That was nice of him.

> *"Prior to the final run, the captain of the ship asked how many crewmembers we had and was given an answer. He said to be sure to open the final capsule. It had a case of cold Pepsi rather than the concrete blocks in the others. That was nice of him."*

The C-130A at the Salton Sea in 1964, probably during a training recovery for a telemetry ship capsule.
/Photo credit: USAF

The Ashcan missions I flew were from Eielson AFB, and it was quite cold there in winter! I went once in the winter and once in the summer. During an Ashcan TDY, we flew one Ashcan recovery a day, normally for a total of four flights. The Ashcan balloons were scheduled for 80,000 feet, 90,000 feet, 105,000 feet, and 120,000 feet. I'm not exactly sure what was said about the Ashcan project mission at the squadron. Other guys who had previously been to Eielson probably told us about Ashcan. The squadron wasn't much on passing out information.

Prior to an Ashcan launch, you knew the balloon ascent time, and the float time prior to cut down, failing unforeseen circumstances, of course. The Ashcan balloons were released on the ground from a ramp area at Eielson. The helium for the Ashcan balloons was in about five or six large cylinders that resembled the oxygen bottles you'd see in a welder. The Ashcan payload had a tubular frame. It had a couple metal cylinders in it approximately the size of a basketball, megaphone-shaped tubes about 3 feet long and perhaps 3 inches wide at the large or inlet end. It also had some other components that really didn't look like anything that was identifiable.

Only one recovery aircraft was sent out for an Ashcan recovery. We had the JC-130 cocked at launch. We stayed in the immediate vicinity of the airplane and watched the Ashcan balloon ascent. Depending on the upper level winds, you were able to see the balloon all the way up to 120,000 feet. As long as it was visible, we'd just stay on the ground at the airplane. If the Ashcan balloon drifted out of sight, we took off and orbited below it.

There were one or more "squibs" attached to the balloon that could be fired by radio control during the cutting down phrase. I don't know what they actually had at the ground station, but we had a tube-like device that fit into the flare port that had a telephone dial on the inside. If the ground signal didn't work, we had a numeric code that could be dialed in to explode the squibs. This ruptured the balloon and initiated the descent of the package. The Ashcan payload was released from the balloon with the parachutes deployed.

The primary difference between Ashcan and Corona, was the Ashcan payloads were suspended by two chutes, both were similar to the Corona chutes. The upper chute was suspended below the Ashcan balloon, then connected to a 150-foot load line, which in turn was connected to the lower chute attached to the Ashcan. At some point in the descent, the upper chute deployed the lower chute, and it descended with both chutes deployed.

You'd try to make a "look-see" (a visual inspection of the descending parachute) at 15,000 feet, fly a teardrop pattern and try to recover it on the next pass. A teardrop is sort of like flying a clover leaf pattern. The pilot went outbound from abeam of the capsule for twenty to thirty seconds, flying on instruments to maintain the same descent rate as the chute. Flying beside the chute, off the left wing tip, we made a turn to the left in order to line up with the chute. Individual pilots preferred twenty, twenty-five, or thirty seconds to give him the time he felt he needed to align with the chute. At his desired time, the copilot called out his time. A turn was initiated, and normally the flight engineer was the first to see the parachute. Each pilot had his own "marker" (the corner of a window or whatever) where he knew he should see the parachute at his desired point in the turn, like flying half of a figure eight. Recovery was made by catching the upper chute. Ashcan payloads were about 4 feet cubed, and in the 200-pound range, perhaps a bit more. Two people could easily handle it on the ramp to position and tie it down.

We were orbiting under an Ashcan balloon on one mission, and it just disappeared! Since the balloon was pretty much directly overhead, you couldn't keep your eyeballs on it 100 percent of the time. Between looks, it just disappeared. The assumption was that it hit some sort of wind shear. We never did see the package descend, and as far as I know, it was lost.

We made only one recovery per flight. Other than training missions using inert bombs (which could be hung from the dolly) or ship data pickups (with canisters light enough to be carried to the front of the cargo compartment), once you had a recovered object stowed on the floor, it was impossible to do more recoveries. I was probably involved in seven Ashcan recoveries. I went up to Eielson twice and one disappeared on us.

As I recall, there were to be six missions. (The NASA BioSatellite project had its first launch in 1966). There were multiple experiments on the capsule, which weighed about 300 pounds I believe. The one project we recovery troops were briefed on involved some sort of radiation exposure and the other was frog eggs. The radiation source was shielded until orbit, and then was supposed to be "re-shielded" during descent. We were given some radiation detectors to check the capsule as soon as it was on board the plane.

The frog eggs' mission was the most interesting, as far as I was concerned. I know nothing of biology, but they told us the eggs have a top and a bottom. In nature, if something causes them to turn upside down, they will attempt to right themselves. If they're unable to do it, and the frogs are born while upside down, they will be deformed. The scientists wanted to see what would happen to the upside down eggs in a no-gravity situation. Personally, I thought, "Who cares?" I suppose we'll never know. The capsule didn't reenter properly. It stayed up until the orbit decayed, and it landed somewhere in Australia. The last I heard, it was never found. We never flew anymore missions for that program, at least while I was there.

Bell gives his final thoughts on the 6593d.

The loadmasters and winch operators assigned to the 6593d had three-year tours, and some (not many) volunteered to extend to a fourth year. I was one of the four-year guys. Being assigned to satellite capsule recovery was and wasn't a good duty for a loadmaster. The young guys up to E-4, and some E-5s, knew they were there as riggers and would remain that until they left. It was a comparatively simple job, all repetitious. Those that had flown the line in a MATS outfit were unhappy at never going anywhere. To the older troops (who had had years of flying the line) it was nice to stay home.

Chapter 7

Maj Gen Donald G. Hard

Maj Gen Donald G. Hard as a brigadier general.
/Photo credit: USAF

"The recovery of Corona 145 was pretty straightforward, really. We had a good hit on the parachute… There was nothing unusual about the recovery at all."

Maj Gen Hard was interviewed on September 28, 2004. He had about 1,500 hours of C-130 flight time when he was assigned to the 6593d Test Squadron (Special) in November of 1969. As a captain, Hard piloted the last aerial recovery of the Corona program when his JC-130 recovered the Corona 145 space capsule on May 31, 1972. He was also assigned to the 6593d when the squadron was deactivated on July 1, 1972 and its mission and personnel were reassigned to the 6594th Test Group.

Hard was happily assigned to the 6593d in 1969 after his tour in Vietnam. He explains why the squadron was a top pick for C-130 pilots and the high caliber of its personnel.

I worked at Los Angeles Air Force Station until 1966 when I went to pilot training. At that time, the current Space and Missile Systems Center was called Space Systems Division. I was very much aware of the recovery operations. In fact, I wanted to go to Hawaii and fly with the 6593d for my follow-on assignment after I'd been in Vietnam.

"As far as C-130 pilots were concerned, the 6593d was considered a tremendous assignment. First of all, the location wasn't too shabby. Secondly, the mission was high priority."

As far as C-130 pilots were concerned, the 6593d was considered a tremendous assignment. First of all, the location wasn't too shabby. Secondly, the mission was high priority. We had good resources assigned. We had an exceptional mission with a lot of priority and resources. If we said we needed something, and could prove that we needed it, we got it. As a result, we had the opportunity to really do well. The 6593d was considered a plum assignment. It was sought by most C-130 pilots that heard about it.

When I arrived in Hawaii in 1969, the squadron had selective manning (personnel were hand-chosen), not only for the pilots, but for everybody. That included even having the ability to say no and to choose the people we wanted for maintenance slots, or any position, so you were in a unit that was one of the best and we had exceptional people there. We only selected pilots with over 1,000 hours of flying time, who were therefore qualified aircraft commanders.

We were the 6593d Test Squadron (Special), emphasis on the word "special." I'll give you an example of what that word "special" meant. Everybody had an additional duty. My additional duty was to be the officer in charge of the Electronic Direction Finding (EDF) and telemetry section. I was amazed when I got there. We didn't do anything the regular Air Force way. If we just wanted some bench stock of 5/8-inch nuts and bolts, we'd call somebody in Sacramento and have them airmail, special delivery, the darn things over to Hawaii. We actually did this, instead of just going through normal Air Force procedures, approaches, and operations.

There was a great turnaround underway when I got there in 1969. By then, we realized that our business was going to require a lot more qualified recovery aircraft commanders, and therefore a lot more training. We started to adopt a lot more of the normal Air Force operations when we better understood that our mission requirements were going to increase dramatically. We had some serious goals.

As far as mission briefings were concerned, Hard notes that they were not told much about the program. In fact, the names "Discoverer" or "Corona" were not used by the squadron. He explains their naming convention.

It was our standard pitch that we were supporting classified Western Test Range activities. That's about all that we got in terms of an in-briefing. There was certainly no briefing on the Corona program, per se. The intent was to declassify as much as possible about the actual recovery operations, because obviously, they were observable, so it didn't make a whole lot of sense to classify them. I knew that Corona was used for reconnaissance purposes, but only because of my previous assignment at the Space Systems Division. I was also aware of the code name "Corona."

It's hard to believe, perhaps, but we never used the spacecraft program names, except in restricted communications (usually classified) in the Mission Planning Shop. The flight crews usually referred to the mission by the type of parachute it used, and we identified the programs simply as "Mission A, B, C, D," and etc in our classified operations plan. When we posted our flying training schedule, it would only show the type of parachute used, such as Mark 5, Mark 8, and etc. Typical phrases were "lights" and "heavies" or "conicals." Our crews were briefed that we were supporting classified Western Test Range activities. Only a handful of people in the group staff were briefed on the actual mission or even the unclassified Air Force program number. So we didn't distinguish between Discoverer and Corona, because we never used those names, except for a very few people in the group planning staff who had to talk with the folks back in Los Angeles and Sunnyvale.

Hard describes copilot recovery training. After pilots were qualified for the aircraft commander status, they still had to wait before moving "over to the left seat." Again, Hard pressed the point that the 6593d had "exceptional pilots."

When I arrived at Hickam Air Force Base (AFB), the pilots came after having a tour somewhere else. The recovery group had selective manning privileges, so they only asked for pilots who had already been checked-out as aircraft commanders. I came as a checked-out aircraft commander with about 1,500 hours of C-130 flying time.

Chapter 7 - Maj Gen Donald G. Hard

JC-130 recovering a parachute and payload./Photo credit: USAF

Once a pilot arrived at the Test Squadron, although you were checked out to fly the airplane as an aircraft commander, you were typically assigned to be a recovery-qualified copilot for a long period of time. You waited your turn for upgrading to aircraft commander. For instance, I arrived there in November of 1969, and although I was qualified for the recovery aircraft commander position several months before, I had my first recovery in May of 1972. That gives you an idea about how long you were in the right seat before you were finally transferred over to the left seat, the Recovery Aircraft Commander (RAC, as we called it) position. That was quite a waiting process. Senior pilots who arrived with 3,000 or 4,000 hours would go to the head of the line, but the junior pilots had to wait their turn.

The 6593d had its own flight training manual. I think we called it the 60-1 in those days. Headquarters Air Force published the Air Force Manual (AFM) 60-1, and it was typically supplemented at the major command level, and it could have been supplemented at lower levels as well. I seem to recall that we had a Test Group supplement that was mostly oriented to our unique mission. If it wasn't an official supplement to the AFM 60-1; it was at least used as such.

The training for the copilot position was typically done in about five or six flights. You were taught all the elements of flying recoveries, and the specific duties for the right seat copilots' job. Familiarization with the overall recovery system included both the parachute and the capsule we were recovering, and also the aerial recovery system and equipment we had in the back of the airplane. A good copilot would spend a fair amount of time learning a lot more than he needed just to fly in the right seat and run the checklist. But that was the extent of the training you received for the copilot position.

The training was much more extensive for the upgrade to the left seat. We spent about six months for this training. I can't remember the exact number of training recoveries we did, but it was probably close to 100 by the time you were through the program. You had a series of lesson plans and a series of check rides. Once you passed the final check ride and were certified to fly in a mission lineup, it was usually worthy of a pretty good party to celebrate. It was a very arduous training regimen that we went through.

> *"You had a series of lesson plans and a series of check rides. Once you passed the final check ride and were certified to fly in a mission lineup, it was usually worthy of a pretty good party to celebrate. It was a very arduous training regimen that we went through."*

I might add, not everybody who qualified as a recovery aircraft commander actually flew recovery missions. We had many pilots who went through the recovery aircraft commander training and, for whatever reason, did not end up in a mission lineup. Each of us who went through recovery aircraft commander training was assigned a RAC number. My RAC number was 104. As I recall, when you looked

Flight crew of the 6593d with a JC-130 in 1962./Photo credit: USAF

at the number of RACs who actually made mission recoveries, about a fifth of them before me had not made a live mission recovery, even though they'd gone through the training. Thus, there had only been about 80 RACs who made live mission recoveries by May of 1972.

I was in the flight scheduling business for a while. When I arrived, we were only keeping about seven to eight RACs current, and we would have one or two in the upgrade training program. We did all that with about 150 monthly training recoveries total. When we later decided to significantly increase the number of qualified RACs, we started off by doubling that to 300 recoveries a month total, and went on to increase that even further, to as many as 350 to 400, depending on whether we were in a mission mode and how many upgrading RACs we had in training. That's an indication of how many pilots we were putting through the RAC upgrade program.

We had exceptional pilots in the 6593d. I recall once we got into a little discussion with some guys from the rescue outfit about who had better qualified pilots. I computed the average flying hours for all of our assigned C-130 pilots, and I think it was well over 3,000 hours, which is pretty good for a group. So, we had excellent pilots with a lot of flying experience over there. Every person there was expected to be a little bit above average anyhow, and wanted to leave the organization even better than he or she had found it, and it showed.

As an aircraft commander, Hard explains in great detail how a pilot successfully flew a parachute recovery.

The flying was pretty straightforward. We used a standard set of maneuvers. Let me start from the beginning. First of all you had to find the "system," as we called the descending parachute and capsule. We had EDF equipment onboard the airplane, which was basically a needle that pointed at the transmitter on the reentry vehicle, much the same way it would point to a transmitter on the ground. It pointed you towards the system, and then you flew towards it and looked for it. If you didn't have a visual indication, if you didn't see the system, you would offset a little bit as you got closer to it, because we didn't like to fly underneath the system. We didn't want to be surprised by not being far enough underneath it. So you offset a little bit, so the needle that pointed towards the system would actually swing when you went by the parachute and capsule. When the parachute and capsule went past your wingtip, then you knew where you were relative to the system. We had procedures for slowing up and flying a box pattern around the system until we saw it.

Once we saw the system, we switched to a different kind of flight pattern, a tear-drop pattern. Basically, you'd fly by the system to ensure you were at the right descent rate, and to visually inspect the parachute and capsule. Then you'd fly outbound for twenty seconds, make a standard rate turn and fly back in, so that you'd have about a twenty-second inbound leg as you went in for the recovery. The basic idea was to fly directly at the system. You tried to descend at the same rate as the system, and then when you turned in on it, you just stopped the system from any apparent right or left movement on the horizon. Then you'd fly directly to and slightly above the parachute. I think we were shooting for an optimum separation of 7 feet between the top of the parachute and the bottom of the airplane, directly under the aircraft's centerline.

In the last five or ten seconds before contacting the parachute, you had an opportunity for fairly aggressive maneuvering by using the rudders to swing the tail over to the system. For instance, if you saw the parachute was going towards the right wing, you could use the left rudder to swing the tail of the airplane over a little bit. Or if you felt that you were loose, or too far above the parachute, you could pull back on the yoke and the tail would drop down a little bit. Sometimes you had to push the nose down a little bit too if you were a little tight in those last few seconds. Sometimes some last minute maneuvers were required, but typically we did pretty well in putting the airplane where we wanted it.

Hard explains the extensive training he and the squadron pilots had to complete. He describes his methodology on how to recover a capsule. Hard adds a personal anecdote on one particular training mission.

Sometimes one airplane dropped training parachutes for another aircraft to recover. However, I would say about 80 percent of the training was with self-dropped parachutes, especially for the lighter systems. When you made a recovery, you would typically be descending at 15,000 feet. The parachute was descending at about 1,500 feet per minute. By the time it got down to sea level, it would be descending at about 1,000 feet a minute or even less.

We flew all of our recovery approaches at 120 to 130 knots. We tried to stay at the low end of that range for two reasons. One was the parachute could actually descend faster than the C-130, so you always wanted to be able to trade increased airspeed for a faster rate of descent if you needed it. Two was, the slower the airspeed, the less the impact, or dynamic force on the system. If you happened to be in one of those red zones, it was more likely you'd be able to catch the parachute, even though it might be in an unsafe area for the hit. You could also adjust the flaps a little bit to vary your descent rate, and we did that too.

We had a couple of different systems by the time we closed down the 6593d Test Squadron and operated as the 6594th Test Group. The light systems, which included Corona, were about 200 pounds. We used a light nylon rope on the small side of the winch. I think it was called "Goldline" rope. We used a light nylon rope on the small side of the winch. We also caught heavier systems. For those we used quarter-inch cable that was covered with a Teflon material. We actually did some 3,000-pound test recoveries, but we typically caught systems that were in the 1,200-pound class.

For the heavier systems it was a little more cumbersome to do a self-drop recovery, so we dropped to each other more often. For the light systems, you could self-drop at 15,000 feet, come back around, and make several practice passes before finally catching it. For the heavyweight systems, you would typically drop it at higher than 15,000 feet because it took a longer time to open, and therefore used up more altitude before you could start working passes. You might get up to 17,000 or 18,000 feet, or somebody else might drop them for you.

We had exercises where part of the training was to find the system. We would drop a system with a training beacon on it at one end of the training area. Then the airplane that was going to recover it, oftentimes with a student, would be on the far end. The task was to come and find it and then recover it. Of course, in order to make it a little more challenging we would always try to find some clouds in between the dropping aircraft and the recovering aircraft. We had a lot of fun with that. It wasn't unusual to intentionally drop it into a layer of clouds, so the student would have to catch the system underneath a cloud. We used to test ourselves pretty hard.

> *"Of course, in order to make it a little more challenging we would always try to find some clouds in between the dropping aircraft and the recovering aircraft. We had a lot of fun with that. It wasn't unusual to intentionally drop it into a layer of clouds, so the student would have to catch the system underneath a cloud. We used to test ourselves pretty hard."*

We had to get one recovery a month above 15,000 feet, and one recovery a month below 500 feet. The true airspeed is much higher at 15,000 feet, so the approach to the system was more demanding. We weren't supposed to catch the parachutes too low because you didn't want the training capsule dragging in the water and making it unsafe for the airplane. There were a few stories about some of the training weights coming back a little bit wet. Sometimes that was connected with an informal test of one's ability before the check ride to make sure that the student was capable of catching the parachute, wherever he had to. That's pretty good flying.

We had a few pilots who would occasionally over-control the airplane. They'd work too hard on the controls and put in corrections that they didn't need, and maybe over-correcting. I remember one of my favorite guys over there, Lt Col Edward "Ted" Lynch who was a flight commander, then an operations officer, and then later the squadron commander (October 31, 1971 to July 1, 1972). In fact, Ted was the last squadron commander before the squadron was closed down and integrated into the group.

I recall Ted giving a particular pilot some instruction and pointing out to him that he could fly hands-off if he had everything really in place. To demonstrate that, Ted went past the system to ensure he had the right vertical velocity, and he kept it exactly the same as the system's during his tear-drop pattern. He turned inbound and put the system on the horizon so the parachute wasn't moving a bit to the left or right. Everything seemed to be lined up perfectly as soon as he had established the airplane for the inbound approach, so Ted said to this pilot, "Now, let's see how much correcting we need here." He took his hands off the controls, both throttle and yoke. About ten seconds out, he gently pushed the yoke just a little bit forward, and we made a very good recovery out of it. I said to myself, "OK, there's something to be learned here." It was all about trimming the airplane.

Going back to the question— "Did recovery get routine?"—to an extent it did if you happened to be lucky and had a good stable chute and a good flying airplane. Some of the airplanes were not as true as others. Some flew with a little bit of a yaw, or the engines and props wouldn't go to idle evenly, so you had a little bit of stuff that you had to work there.

There were multiple types of parachutes used by the 6593d during recoveries. Some were used for lighter-weighted capsules, while others were for payloads weighing up to 3,000 pounds. He discusses some of the challenges of each.

"The more you flew the easier it became, but everyday was a challenge."

The more you flew the easier it became, but everyday was a challenge. Even if you were an old head with a couple of years and several hundred recoveries behind you, you could still get surprised on any particular day. It was also a function of the parachute. One of the things that we had to do at the test squadron was transition from one recovery system and parachute to another. There were very different sight pictures in terms of what the parachute looked like and how it behaved during a recovery. Because of that, we typically liked to keep our mission crews on one system, rather than swinging them back and forth between two systems, but were not always able to.

All of our training chutes were refurbished. In fact, we spent good money improving the parachute facility at Hickam AFB to ensure they could handle the parachute refurbishment that was required to support our training.

JC-130 about to recover the white cone (or conical extension) of a Mark 8 parachute. The cone had reinforcement lines that were snagged by the JC-130 hooks during a recovery./Photo credit: USAF

JC-130 featured in a photo from the Air Force Systems Command, Hickam AFB, Hawaii./Photo credit: USAF

The Mark 5 training chute for the small Corona-type buckets was really a squirrelly parachute. It did a lot of crazy things. It kept you on your toes. You really had to be quick and agile because the Mark 5 would move around a lot. If a parachute was refurbished perfectly symmetrical, it was pretty stable. But occasionally it flew a little squirrelly for whatever reason, especially after it was recovered and refurbished three or four times. Sometimes the chute had an airfoil shape to it, and it would actually fly in one direction or the other, so you had to be prepared for that. Sometimes they would open and close. That pumping kind of action would change the descent rate. It could be very demanding. In the case of the Mark 5, you had to be just as careful on your 250th recovery as you were on your second recovery because of the parachute itself.

The Mark 8 was a bigger parachute for these larger loads. From the top of that parachute to the bottom was about 100 feet, which is almost the length of the C-130. The Mark 8 was a standard parachute but it was huge. It had a cone on top of the parachute, and the cone is what we aimed for and hopefully contacted, not the main parachute. In fact, we avoided the main parachute. If we were unsuccessful in our recovery attempt and tore open the cone, the main parachute would let the capsule descend gently into the ocean so it could be recovered by helicopters. The Mark 8 cone was about 20 feet from bottom to top, and about 15 feet across, so you had a little different sight picture. Of course, for a recovery you were trying to put the loop about one third of the way up from the bottom of the cone, and trying not to get the loop into the main parachute. It was a little different than recovering the Mark 5. The Mark 8 training chute, for whatever reason, was boringly stable. We used to do all manner of things to try to get it to move around a little bit, because the mission chute was quite the opposite. We seldom had to take a second pass on the Mark 5 mission chute.

I was the first pilot who had to take three passes to recover a Mark 8 on a live mission, with two intentional attempts to make a recovery. In one case the parachute flew to the side and there wasn't any way I could move to it. In the other case, the parachute just buckled up and dropped and there wasn't any way I could get to it. You could get surprised on mission day too.

The nylon recovery line had a lot of elasticity in it. Sometimes if you tore through a parachute, or something else happened, the nylon line was like a rubber band. When the snagged parachute broke loose, the force of the stretched nylon line came back into the airplane. Sometimes it came back in the open rear door, and sometimes it got caught in the empennage. I can recall bringing an airplane back once with the loop and hooks caught up on top of the vertical fin. The hooks ripped the skin of the fin all way down to its base, just like you unroll the metal on a sardine can.

We had another parachute system that we operated with, not for very long, called a tandem parachute. It had a Mark 5 parachute below, and then it had a 30-foot line up to another little parachute, which was the pickup chute. That line between the pickup chute and the main parachute was a squirrelly thing. We had several cases where those parachutes went into propellers or worse. We had several cases where we'd get a little too tight on the parachute during a recovery. We'd clean off the bottom of the airplane when a parachute brushed against it. We used to brag that we had the cleanest C-130s anywhere in the fleet because we'd literally taken everything off the bottom of the airplane.

> *"We'd clean off the bottom of the airplane when a parachute brushed against it. We used to brag that we had the cleanest C-130s anywhere in the fleet because we'd literally taken everything off the bottom of the airplane."*

We had some wonderful film back at Edwards AFB of a recovery attempt on a new experimental parachute that was huge. I think it was called a Mark 9 parachute. You could see the recovery pass from the nose camera film. The airplane was very, very tight, clearly too tight. That was just before the screen went international orange as the parachute covered the nose and cockpit window. They had to literally cut holes in the parachute through the cockpit windows to make enough room to see and to land on the lakebed.

Hard describes the competitive dynamics between the recovery pilots. He goes into specifics about the "RAC Rater" competition.

Our airplanes had a nose camera system installed in them. The camera monitored the last ten to twenty seconds of the inbound recovery run, so you could later see the system as you approached it. We also had high-speed cameras that recorded the way the parachute contacted the recovery hooks and loop that we had trailing behind the airplane. Lou Bedwell, a Lockheed employee, took all that recovery film and projected it in such a way that he could score the recoveries of each pilot. Then that score was plotted, so you generally knew how accurate each pilot was.

This was pretty impressive flying, really. I'm not bragging about myself, but all the pilots who flew space capsule recoveries. If you made a perfect recovery contact and hit the parachute exactly where you were supposed to, within the poles and the loop behind the airplane, then that was called a zero-point recovery. This was when the reference point on the skirt of the parachute contacted the loop exactly where you wanted it. If the recovery contact had been 1 foot loose or high on the parachute, it would have been scored as 1 foot off in the vertical, so it was a one-point recovery. If it was 1 foot loose and 3 feet to the side, we didn't bother to do the square roots or anything, we'd just add those coordinates. So, if you had a recovery that was 1 foot loose and 3 feet to the right it would be one plus three, or a four-point recovery.

Typically, a pilot would make fifteen training recoveries a month, unless he was in a mission mode, then he flew more recoveries to ensure that he was ready for the mission. We scored all the recoveries, but we took the scores of the first six recoveries of every pilot and used them in a competition that we called the "RAC Rater." The pilot with the best score for those six recoveries would be the RAC of the month. The pilot with the worst score would get severe humiliation. We were organized in three flights when we started, and then we went to four flights. The flight that did the worst in the recovery scoring, or the RAC Rater competition, would buy the beer for a monthly party. So, there was a little incentive to improve your scores.

Typically, the winning score was around fifteen points for the six recoveries. That is about a 2.5-foot average, or about a 3-foot circular error of probability for recovery accuracy. That ain't too shabby! I think the best recovery score we had was eleven points for six recoveries by Capt Glen Messerli. I recall, he had two perfect recoveries in the six that were scored.

If the parachute was missed completely, it could still be scored if you could see it on the film. There was a penalty, though. I think an additional five penalty points were added to your score if you had what we called a "red zone" hit. The zones for recoveries were set up so that if the parachute apex was outside the pole reference point, then it was unsafe. If it was below a certain point, or too tight, it was also scored as an unsafe recovery. Even though you made a successful recovery in those areas, you still got five points added to your score if it was in the red zone. I think if it was more than 13 feet below the desired elevation, it was unsafe. So, if you had been perfectly lined up at 14 feet, you'd get fourteen plus five for a total score of nineteen because it was a red zone hit.

JC-130 reeling in a test capsule at Edwards AFB in 1961./Photo credit: USAF

There was an additional penalty (I believe twenty-five) for parachute tear-throughs. We used the term "tear-through" to describe an unsuccessful recovery attempt when the airplane's recovery loop just tore through the parachute's silk and you didn't bring the system onboard. There was more humiliation that came along with that. We had this trophy that captured the results of a tear-through. It was a set of old knotted-up loop and silk from torn parachutes on a battered up old plaque, just an ugly thing. Whenever you had a tear-through, that trophy came to your flight room door and you owned it. We called that trophy the "rag." You kept it until somebody else was unfortunate and had a tear-through, and then you passed it on the next unlucky pilot.

I can recall when we were setting up the rules for the RAC Rater competition. Lt Col Bill Scott was one of the "old guys" (pilots who had 5,000 to 6,000 hours in the airplane) and he was our best pilot. The old guys were excellent pilots, marvelous pilots. Bill was opposed to the initial rules that competed with just one recovery rather than six, making the case that anybody with one perfect recovery could win the competition. Bill was discussing it with Maj Jack Swatek, another hero in the squadron, who was putting the competition together when we first started it. He said, "Jack you need more than one recovery score. If it's only one recovery, then an old head like me could go out and make one perfect recovery a month to win the RAC Rater contest every time." Jack looked at him and said, "If you can make one perfect recovery a month, I'd like to see that." Bill proceeded to do just that for two months in a row, so we changed the rules. We were very competitive, but the competition was all focused on the RAC Rater competition. It was all about accuracy.

If you looked at all of those training recoveries for certified recovery aircraft commanders (not including initial upgrade training), and scored them all, and took the circular error probability, I would guess that 80 to 90 percent of them were less than 5 feet away from the desired reference point. So we had an airplane with a 130-foot wingspan, and we routinely put it into a 5-foot box or a circle. That's less than 5 percent of the wingspan. Again, that's pretty good flying!

Hard explains the procedures for conducting live recovery missions. Each aircraft had a position number. He details what each position entailed. The overall mission day was well orchestrated with all of the personnel having a clear picture of what was expected.

We intentionally moved our crewmembers around, except for mission line-ups. When it came to a mission, the crews that were assembled for that mission operated as an integral crew during all the training and alert days for it. When we were in mission mode, which was whenever the host satellite had reentry vehicles on-orbit, we were prepared to conduct a recovery each day until we were released for that day.

We had the full five-crew/five airplane normal force that was prepared to fly if it needed to. Then we had a three-airplane and two-crew emergency force that was kept on alert until the last possible minute, so it could have gone and made a recovery as well. Those crews were integral for the duration of the mission, but when they were put together for normal training, the crews were put together based on training needs. As we got towards the tail end of the business, we realized that we were getting better in our precision, so we could conduct the missions with fewer aircraft, in fact for one program that we supported, we dropped down to only three required aircraft.

We typically assigned mission crews out of the same flight, but even there, once in a while, we'd end up with somebody from another flight. The navigators weren't assigned to the flights. They just had one flight of navigators, so that was a little different. We tried to match up experienced folks with inexperienced folks in all the positions as best we could. There was no intent to try to keep the crew intact.

On mission day, the number two aircraft flew as a backup position on the primary aircraft, and it would have onboard an on-scene commander, or "lead" as we called them. The airborne lead was typically a senior RAC who had made several recoveries himself. He was typically a flight commander (a major or a lieutenant colonel) and someone who had the maturity, the experience, and the hands-on ability to conduct any necessary changes to the planned operation out at the recovery area. The airborne lead was just, more or less, a referee and an airborne senior guy in charge out there in case there was any question about what to do next.

We would get a Projected Impact Point (PIP, as we called it). From that, there was a dispersion projection, if you want to look at the engineering side of it, which we tried to cover based on the ability of the C-130 to run a particular distance and get to the system in time to make a recovery. If your projected impact point was here, on a track of the satellite, then the probability of catching it was in the ballpark where we would have orbital traces to cover. The ballpark was the area around the Hawaiian Islands that you could reasonably reach with the C-130. Depending on the PIP location, we would establish the areas of actual coverage for a possible aerial recovery. These tracks were about 60 by 200 miles. We would have five airplanes stationed to cover the areas indicated. Number five was down low in case the parachute and capsule went long. One of these aircraft would carry PJs, specially qualified rescue personnel, to jump from the airplane into the ocean if the reentry capsule ended up in the water and they needed to secure it for a surface recovery by helicopters.

H-3 helicopter on the deck of a recovery ship supporting the Corona Program./Photo credit: USAF

On mission day, our operational procedures called for the number one aircraft, the designated primary aircraft, to make the recovery. The number two aircraft would also be close to the PIP. Number two was the backup position on the primary aircraft. Number three's job was to fly below the lowest cloud deck to be able to make a recovery down there. That's the way we lined the aircraft up. If something happened where another aircraft was better positioned and ready for the recovery, and it was apparent that the lead aircraft wasn't going to get to the system for quite some time, I forget the exact rules, but then you could designate another aircraft as the primary aircraft. The point is, with five airplanes, we would typically have several opportunities for a recovery. It was the airborne lead's job to identify who was going make the next pass.

Typically, you'd get to the point underneath the system well before the parachute would get down to a working altitude. The parachute opened up at 55,000 feet and descended at 1,500 to 2,000 feet per minute above the 15,000-foot altitude. That gave you plenty of time to get there and wait for it to come to your altitude. We typically held our position there at 20,000 feet or higher, flew where we expected the parachute to come down based on the EDF equipment and waited there. Then we descended to 18,000 to 15,000 feet, and got the airplane in the right configuration by deploying our recovery loops. Presuming that you started recoveries right at 15,000 feet, any aircraft by itself would have at least four or five passes at the system before it splashed in the sea. With three aircraft, there would typically be time to easily make ten or fifteen passes altogether, depending on the weather. We never used more than three, and typically used only one or two. So, we had a lot more opportunities for the recovery than we needed, if the system was anywhere near its nominal impact point.

If the reentry capsule went long or went somewhere else, then the first airplane on the scene automatically became the primary. We had a few capsules that descended in unexpected areas, more of them in the earlier days than in the later days. We had them marked off in grids.

Only once in my tour there do I recall one that got recovered out of sequence. Capt Marshall Eto (who caught the last bucket for the test group in 1985) was designated to be the primary aircraft commander by the airborne lead because he happened to get to the system first, and was the only one there ready for a recovery. That was before I was checked out as a recovery aircraft commander. I think in every case that I can recall after that, the primary aircraft made the recovery.

There was not much public celebrating after a capsule recovery. Hard says there was not much they could do except to celebrate privately due to security concerns. After he recovered the last Discoverer capsule, #145, the celebration protocol hadn't changed. Hard gives his final thoughts on both the 6593d and the 6594th.

We had a very special procedure because of the security involved. After a recovery, on our way back to Hickam, we would put the capsule into a shipping container that was specially built for it, and locked it up. After we landed at Hickam there was a transfer of papers and signatures. Then the container with the capsule was loaded into one of our recovery trucks and taken to another aircraft for transportation back to the continental United States. A different unit provided the aircraft that transported the capsule from Hawaii, typically flying a C-141 aircraft. We didn't have anything to do with that part of the operation.

The aircraft commander had to buy drinks at the bar when we went back. There wasn't any official recognition. We had parties after each of these missions, and the successful aircraft commander was expected to buy at least the first round or two of drinks in two locations, one at the officers' club and one at the enlisted club.

By the time I left, we had what we called "aces," pilots who made five recoveries or more, and we had special parties for them. There was always a big party for a pilot who made his first mission recovery. That whole party was at his expense, you're talking about several hundred dollars back in the 1970s when things weren't all that cheap. A successful recovery mission was a pretty big deal, but because of the nature of the mission, our intent was to keep it kind of low-key due to the security considerations, so there wasn't a lot of recognition other than a recovery party. At the end of your tour, maybe you got an Air Medal if you did real well, but that was about it.

> *"A successful recovery mission was a pretty big deal, but because of the nature of the mission, our intent was to keep it kind of low-key due to the security considerations..."*

In the case of the last Corona recovery, this was my first mission as the number one RAC. Prior to that, I probably flew less than a dozen live missions as a RAC with all the supported programs, not just Corona. That was basically the number of missions I supported while waiting for my turn to fly the number one aircraft. As a copilot, I probably flew three times that many live missions.

Although the flight engineers did a detailed walk around the airplane, the pilots would also typically do one with the crew chief, to some lesser extent. I recall going out to the airplane and my flight engineer said, "Just get into the cockpit and we'll go fly." That was an indication that I shouldn't look too hard at the airplane. So I didn't and we took off.

The recovery of Corona 145 was pretty routine, except that we had bad weather in the ballpark. The ballpark would always be cut by at least one trace and oftentimes two traces of the satellite each day, so you'd have at least one opportunity and sometimes two, to recover it. We had lousy weather under both traces, so we weren't really sure where the best area would be to bring the system in, so we split the force. Capt Don Krump went with four airplanes to the west and I went with four airplanes to the east. When you did that, you had to fly all the airplanes. I think the 6593d only had thirteen C-130s assigned, and we typically had one or two that were in the Continental United States for scheduled maintenance or refurbishing, so it was kind of hard to generate ten airplanes. You'd have ups and downs in maintenance too.

The recovery of Corona 145 was pretty straightforward, really. We had a good hit on the parachute. Typically, you'd tend to make contact with the parachute a little bit tighter on a mission recovery, and I did. It was a good square, solid, and safe hit. We recovered it on the east side. It had better weather there. As soon as we got the system onboard, we shut down the number two engine because of a prop leak that my flight engineer had correctly assumed was a very slow leak. We could (and did) live with the leak, but by the book, had I seen it, we probably should have refused the airplane or shut it down earlier, and that would have taken me out of the primary pilot slot.

There was nothing unusual about the recovery at all, except for the east side-west side approach, and having to shut an engine down, but it wasn't that unusual. Flying with three engines in the C-130 is not a big deal. You fly the C-130 the same way with three engines as you do with four engines. You might take a few precautions if it's an engine on the left side, which powers some of your hydraulics, but typically there's not that much difference in the airplane handling characteristics. So, it's not like it's an emergency to shut one down. The airplane flew pretty decently with three engines. Anyhow, we flew back to Hickam uneventfully. I knew at that time that this was the last Corona recovery. There wasn't any ceremony for Corona, because we couldn't do it publicly.

I wasn't in on the decision process. I'm pretty sure the 6593d moved to the group structure rather than the squadron structure mostly for efficiency and to save a few manpower spaces. At the same time, we had another squadron in the group that was also deactivated. It was an instrumentation squadron that later become known as the Hawaii Tracking Station, and it still operates under that name. I don't think the instrumentation squadron was attached to a particular mission, because we deactivated the squadron at the same time that our requirements for recovery were going up, not down. I don't think the deactivation was associated with the end of the Corona program.

I don't even think we had a deactivation ceremony for the 6593d. If we did, I wasn't there. I don't recall anything. I'm sure there must have been something official, but I don't recall it. On the other hand, I certainly recall the deactivation ceremony for the 6594th Test Group in 1986. Going back to command the 6594th Test Group was the best tour of my whole thirty-one-year Air Force career. It was a very nostalgic and sentimental return and departure for me when I left. I didn't want to leave. It was the best.

> *"Going back to command the 6594th Test Group was the best tour of my whole thirty-one-year Air Force career. It was a very nostalgic and sentimental return and departure for me when I left. I didn't want to leave. It was the best."*

Chapter 8

Col Harlan L. Gurney

Col Harlan "Bud" Gurney in 1974
/Photo credit: Harlan Gurney

"I could imagine the headlines if the system were to land in the streets of San Jose…"

Col Gurney (1928-) was interviewed through e-mails between July and August 2003. He was a C-130 Recovery Aircraft Commander (RAC) for both Operating Location Number 1 (OL-1) flight testing and the 6593d Test Squadron (Special) recovery operations. He piloted the recoveries of Ashcan packages and three Corona space capsules (Corona 116 and 121 in 1967 and Corona 126 in 1968). A limited amount has been written with first-hand perspectives of Corona recoveries, and much less has been published about the aerial recovery of other projects. Col Gurney describes the development of the aerial recovery mission for the Ashcan project and its recovery operations.

Col Gurney arrived at Edwards Air Force Base (AFB), California, in 1963. He was checked out and qualified as a recovery pilot. Gurney moved to Hickam AFB, Hawaii, in 1965.

In early 1963, I was assigned to the 6594th Aerospace Test Wing, OL-1, at Edwards AFB. It was the research and development test detachment for aerial recovery equipment, parachutes, and techniques. I was quickly checked out in the C-130 recovery aircraft and became qualified as an aerial recovery pilot while flying aerial recovery equipment and parachute tests. In 1965, I was assigned to the 6593d Test Squadron (Special) at Hickam AFB.

Gurney was directly involved with the development of the Ashcan recovery system. He recounts the program and the recovery requirements he had to work with. After much testing, they had a proper recovery system.

In late 1963 or early 1964, the high altitude balloon section of the Air Weather Service (AWS) at Goodfellow AFB, Texas, was tasked to obtain air samples from balloons flying for four to six hours at 90,000, 105,000, or 120,000-foot altitudes. After the samples were collected, the entire sampling assembly would be released upon a radio command and returned to earth by parachute. The sampling assembly was suspended from the balloon, along with a parachute. It consisted of two individual sample collecting blowers, with auxiliary bags containing timers, batteries, a radio transmitter and a command receiver, all spread across a horizontal balloon bar that was 10 or 12 feet long. Two of the desired areas of operation were Panama and

Alaska, where either the wilderness or the ocean prevented a ground party from following the balloon and retrieving the sampler assembly after it returned to earth.

Accordingly, the Ashcan project officer from the AWS, a USAF captain, shipped a complete balloon bar sampler assembly to OL-1 at Edwards AFB. He asked that we consider using aerial recovery techniques to recover the Ashcan sampling assembly. The balloon bar configuration was clearly unsuitable for a safe recovery boarding in flight, and the accessory bags were unlikely to withstand the stresses imposed on them during an aerial recovery. Also, the weight of the assembly would result in a descent rate that was too severe to enable a recovery using the available standard Mark 5B, 28-foot nominal diameter, ring-slot recovery parachute. An appropriately sized parachute did not exist then that was reinforced to withstand the forces imposed on it following the hook engagement during a recovery. However, it appeared that aerial recovery might be feasible if the sampler package with its two blowers were reconfigured, and if the existing recovery parachutes could be adapted to accommodate the package weight.

Capt Robert D. Counts, the OL-1 navigator, designed a core rectangular "cage" to contain the required accessory equipment and batteries. He then supervised its fabrication in the huge maintenance and modification hangar at Edwards AFB. The two Ashcan blowers were attached to adjacent sides of this cage to form a compact, 90-degree, V-shaped assembly. Trail test flights were made to optimize a harness for the package and make it as stable as possible in the air stream, and thus safe to winch aboard after a recovery. The reconfigured sampler package proved to be very stable and easy to bring aboard.

It appeared that two Mark 5B parachutes tethered in tandem (one above the other, connected by a suitable lift line) should provide a rate of descent typical for recoveries by the JC-130B aircraft

JC-130 recovering a parachute and payload in 1961 at Edwards AFB./Photo credit: USAF

JC-130 with a parachute and payload./Photo credit: USAF

(approximately 1,500 feet per minute). In test drops, using various tether lengths, it was found that 75 to 100 feet of separation between the upper and lower parachutes was adequate to ensure that the upper parachute, descending in the wake of the lower, was sufficiently stable for safe and consistently successful aerial recoveries. The lift line connecting the parachutes consisted of doubled nylon straps, running from the suspension lines on the top parachute, down through the apex of the bottom parachute, and to the junction of the lower suspension lines with the sampler/blower package harness. The apex of the bottom parachute was pigtailed to the lift line at the location appropriate for the parachute when inflated during descent. The parachutes and the sampler package were deployed in a vertical "string" suspended from the balloon as it was launched. The deployed parachutes opened during their free fall, following the release of the string from the balloon.

The release from the balloon was activated by commands from a Motorola circular telephone, dial-operated, radio command unit. The Motorola set controlled the functions of the balloon sampler package, as well as its release from the balloon, and required an external antenna on the aircraft. Capt Counts supervised the machining of a Bakelite plug to which the antenna was mounted. It fit through the Very pistol (emergency signal gun) port in the top of the aircraft fuselage above the navigator's station. The Motorola command unit sat on the navigator's table, and was powered by an adjacent aircraft electrical power outlet, with a coax (cable) running up to the Very pistol port plug and the antenna. Thus the Motorola unit could be installed in any JC-130B aircraft without the need for aircraft modifications.

The typical Ashcan mission profile for the C-130 recovery aircraft was to take off prior to the balloon launch. The C-130 circled around and under the balloon as it ascended, monitored the Ashcan package telemetry radio signals, and flew orbits below the balloon to keep it in sight as it collected its sample. Some of the balloons drifted a long way, some circled around erratically, and some drifted one way for a while before backing up or going off in another direction. An expected drift track was generated before each balloon launch, a prerequisite for obtaining flight clearances for the recovery aircraft.

The balloons generally drifted as predicted. Sometimes balloons were destroyed by wind shears they encountered that ascended to and through the tropopause. Balloon ascent through the "trop" was always a "cross your fingers" time. When the sampling was completed, depending upon where the balloon had drifted, one could choose the time, and to a limited extent, the location for the package release. One made certain that the balloon and package were in sight when the package release was commanded from the aircraft, to assure continuous visual tracking of the parachute system as it descended to the recovery altitude of 15,000 feet or below. Following a successful aerial recovery, the package would be unloaded after landing. The sample was removed and sent to the laboratory for analysis.

For the Ashcan missions supported by the 6593d Test Squadron in Panama, the balloons were launched across the Panama Canal from Albrook Field, and the recovery aircraft left from Howard AFB. In Alaska, both the balloon and the aircraft were launched from Eielson AFB, south of Fairbanks.

Gurney discusses the Ashcan missions he flew. Not all went according to plan, but he and his crew were able to mitigate the factors and fly successful missions.

We had previously flown one short test mission, or perhaps more, from Holloman AFB, New Mexico. The balloon launch was from Walker AFB, New Mexico, around 1964. The very first deployed, operational Ashcan mission attempted was from Eielson AFB in early November 1964. I flew the first Eielson mission with our Edwards AFB, OL-1 assigned, C-130A model, serial #53-3131. It was equipped with a Model 80 winch, C-119-type recovery gear, and a Benson auxiliary fuel tank in the cargo compartment.

On the first Ashcan mission, it was necessary to de-ice the aircraft with an alcohol wash at the fuel pit in Eielson. On the first take off from the ice-covered runway to support the balloon about to be launched, the Benson tank cap popped open immediately after takeoff, and fuel gushed onto the cargo compartment deck. I declared an emergency, did a 180-degree turn, and landed in the opposite direction from take off. After opening the cargo door and ramp, fire trucks washed down the entire rear of the aircraft. It made quite a mess on the ramp.

We then took off again, and recovered a parachute from a balloon at 90,000 feet, which required a relatively short time at drift altitude. We had another emergency when the pilot's rear side cockpit window failed under pressure at altitude while following under the balloon. For the remainder of the mission, it was necessary to fly unpressurized and go on oxygen as necessary. In the Alaskan cold, our venerable Mojave Desert-acclimated aircraft developed fuel leaks across the entire extent of the wings containing the fuel tanks. We required headquarters' permission for a one-time flight to the depot at Robins AFB for a total IRAN (Inspection and Repair As Necessary) prior to returning from Alaska.

The second mission in Alaska was used to instruct and qualify a 6593d Test Squadron (Special) crew for Ashcan missions. The flight orders designated that I (even though I was assigned to OL-1) would be the RAC on the mission. The JC-130B aircraft, with a complete Hickam crew, including a RAC (Capt Doug Sliger), flew into Edwards where I and OL-1 navigator Capt Robert Counts went aboard, then flew up to Eielson.

This mission used a balloon flying at 120,000 feet, so it drifted at altitude for about six hours getting its sample in the thinner air. The lower parachute did not deploy following its release from the balloon, and the package was not recoverable because of a couple of ancillary problems. It landed near Circle Hot Springs, Alaska, and was subsequently retrieved by an H-21 helicopter. In assembling the original Ashcan parachute system, one half-inch nylon rope had been used for the lift line from the top parachute, through the bottom parachute, to the package. It was suspended for so long a time that the rope had untwisted, twisting up the bottom parachute so that it would not open. The fix was to employ a doubled nylon strap as the lift line thereafter.

It was about the 1964 or early 1965, that we flew the C-130A (#131) from Edwards (via Kelly AFB, Texas) to Howard AFB. I guess that we made one or two recoveries on this trip. I believe that I flew only one Ashcan mission series (three balloon launches and package recoveries) after being reassigned to the 6593d Test Squadron at Hickam. This was in late 1965 or early 1966, again in Panama. Capt Duncan B. Parker, also a RAC, was my copilot, as this was his mission familiarization flight.

I recall one of the recoveries vividly. The Ashcan balloons could drift a long way, and this balloon drifted directly over San Jose (the capital of Costa Rica) and a group of building thunderstorm cells. Telemetry signaled the sample's completion. The release of the package was commanded from the balloon, above what was hoped would remain a small break between cloud buildups. However, the clouds below had thickened into a solid blanket with a low base, and the parachute system drifted towards the side of a towering cumulus. I could imagine the headlines if the system were to land in the streets of San Jose, so with the rig deployed for recovery, we climbed desperately to try to recover the package before it entered the clouds. The Standard Operating Procedure (SOP) for the maximum recovery altitude limit was 15,000 feet, but we made the recovery at 19,800 feet, just as the system was entering the cloud edge. We were immediately bumping along within the cloud, making a 180-degree turn, and descending as we headed back towards the ocean. The package reel-in began after getting clear of the weather. To my knowledge, that remains the altitude record for any aerial recovery. Back at Hickam, I was forgiven for violating SOPs when the recovery photographs were reviewed, showing the weather conditions. It was nice that the aircraft's recovery documentation cameras had functioned properly.

Gurney reflects on his time as an aerial recovery pilot. He compliments his aircrew as well as all of the 6593d squadron.

I was the recovery pilot for every operational Ashcan mission I flew on. In summary, I probably made a number of Ashcan parachute recoveries at Edwards during the Ashcan parachute system and package development/feasibility tests, a recovery from an Ashcan balloon launched at Roswell, one successful and one unsuccessful Ashcan recovery attempt in Alaska, and perhaps four or five total recoveries of systems launched from Panama. This is my best guess, based on my memory of the events that occurred 30 to 40 years ago. There may be omissions as well. I seem to remember flying directly to Eielson from Hickam on one occasion, but do not recall the corroborating details. Throughout my years as a recovery pilot from 1963 to 1971, simultaneous involvement with several programs was the norm. At Hickam we were flying a minimum of six to eight recoveries a month for training, or for operational program support (often more at OL-1 for testing).

Although this interview is a first person account, I would be remiss if I did not mention that successful aerial recovery required a team effort by every member of a highly trained and specialized aircrew of ten persons: pilot, copilot, navigator, flight engineer, electronic systems operator, winch operator, and four riggers. Aerial recovery crews from the 6593d Test Squadron (Special) flew a number of Ashcan mission series, perhaps for several years. I am not able to provide an estimate of how many aerial recoveries were made supporting the Ashcan project, but it was a privilege to have been a participant. In that period of the Cold War, it was a project of national importance.

> *"I am not able to provide an estimate of how many aerial recoveries were made supporting the Ashcan project, but it was a privilege to have been a participant. In that period of the Cold War, it was a project of national importance."*

Chronology of the 6593d Test Squadron (Special)

July 1958—The Air Research and Development Command (ARDC) was directed to form a C-119 unit for the aerial recovery of deorbited space capsules. Tactical Air Command (TAC) was tasked to provide the personnel for the new unit. Representatives from ARDC and TAC met with the Air Force Ballistic Missile Division (AFBMD) officials in Los Angeles to organize the new aerial recovery squadron—the 6593d Test Squadron (Special). Maj Joseph Nellor was selected as the first squadron commander.

Aug 1, 1958—The 6593d Test Squadron (Special) was activated. The squadron personnel would report for Temporary Duty (TDY) at Edwards Air Force Base (AFB), California.

Aug 5, 1958—The commander and an administrative clerk were the first squadron personnel to report to Edwards AFB.

Aug 21, 1958—The squadron's nine C-119J aircraft began arriving at Edwards AFB from the Fairchild Aircraft modification facility in St. Augustine, Florida. The C-119Js continued arriving at Edwards AFB through September 25, 1958.

Dec 2-10, 1958—The squadron transferred its personnel and eight C-119Js (C-119J #51-8050 remained at Edwards to conduct water recovery tests) from Edwards AFB via Travis AFB to the squadron's permanent duty station at Hickam AFB, Hawaii.

Dec 15, 1958—The aircraft and aircrews were in place and operational at Hickam AFB.

Feb 28, 1959—Discoverer 1 was launched from Vandenberg AFB but the Agena failed.

Apr 8, 1959—Nine photographers from the 1365th Aerial Photo Squadron at Orlando AFB, Florida, arrived at Hickam AFB for a 180-day TDY with the 6593d to provide film and photographs for Air Force news releases.

Nov 1, 1959—The squadron was assigned to the 6594th Recovery Control Group at Hickam AFB.

Aug 11, 1960—The Discoverer 13 capsule was floating in the Pacific when it was recovered by a Navy diver and helicopter, the historic first recovery of a space capsule.

Aug 19, 1960—Capt Harold Mitchell piloted the C-119 recovery of the Discoverer 14 capsule, the historic first midair recovery of a space capsule. Discoverer 14 was also the first satellite to provide photoreconnaissance and began a revolutionary intelligence breakthrough.

Jun 1961—Three JC-130Bs were delivered to Hickam AFB by this time as the squadron converted its recovery aircraft from the C-119.

Jul 9, 1961—Capt Jack Wilson piloted the C-119 recovery of Discoverer 26. This was the last Discoverer capsule recovered by a C-119.

Jul 14, 1961—The Air Force approved the official 6593d squadron emblem.

Sep 14, 1961—Capt Warren Schensted piloted the recovery of Discoverer 30. This was the first mission that a JC-130 recovered a Discoverer capsule.

Nov 1961—The Air Force filmed "Catch a Falling Star" at Hickam AFB that featured the C-119 aerial recovery mission.

Jan 25, 1962—The last two C-119s were transferred from the squadron.

Mar 3, 1962—Capt Jack Wilson piloted the recovery of Discoverer 38. It was the last mission known as "Discoverer." The Discoverer designation was changed to Program 162 afterwards.

Jul 21, 1962—The Air Recovery Section (Helicopter) was officially authorized for the squadron. The section had three H-21B helicopters that recovered space capsules from the ocean when a midair recovery was not successful.

Oct 3, 1962—Squadron JC-130s were the primary contingency force to locate the Mercury 8 space capsule in the event that it landed away from its expected recovery area. The aircraft were deployed and tracked the capsule as it descended, but the capsule was recovered in the primary impact area.

Nov 3, 1962—Maj Gen Ben Funk, Space Systems Division (SSD) Commander, participated in an aerial recovery training mission.

Mar 19, 1963—A squadron CH-21 helicopter (51-15872) was lost in a crash at sea during a support mission for the Pacific Air Rescue Service. No injuries or deaths were mentioned in the report.

May 16, 1963—Three squadron JC-130s were the primary airborne force to locate the Mercury 9 space capsule during its descent by tracking it with the aircrafts' electronic direction finding equipment. One JC-130 was staged out of Johnston Island and two were staged out of Midway.

Jun 8-9, 1963—During a visit to Hawaii by President Kennedy, the squadron provided two helicopters for the local rescue capability during the arrival and departure of the presidential aircraft.

Dec 1, 1963—Three CH-3B helicopters were delivered to the squadron as support and rescue aircraft with the additional mission as a backup for the JC-130Bs in recovery.

Feb 22, 1964—Capt Jeremiah Collins piloted the recovery of the second reentry vehicle (RV-2) capsule from Corona 76. This was the first mission when two space capsules were recovered from one Corona satellite.

May 15, 1964—The squadron sent an aircrew with a JC-130 to Eielson AFB, Alaska, to provide midair recovery support for the Ashcan Project. The squadron would provide ongoing support for Ashcan.

Jun 12, 1964—Maj Jack Wilson piloted the recovery of RV-1 on June 8th and RV-2 on June 12th from Corona 79. This was the first dual recovery by the same pilot of both capsules from a single Corona satellite. Program 162 was redesignated to Program 241.

Oct 13, 1964—Capt James Varnadoe received an award from Maj Gen Ben Funk, SSD Commander, for being the first mission qualified aircraft commander to complete 100 consecutive successful aerial recoveries.

Chronology 6593d Test Squadron (Special)

Dec 30, 1964—Capt James Varnadoe piloted the recovery of RV-1 on December 24th and RV-2 on December 30th from Corona 90. This was the second, and final, dual recovery by the same pilot of both capsules from a single Corona satellite.

Feb 1, 1965—The 6594th Recovery Control Group assumed operational control of two surface recovery Victory Ships, *Longview* and *Sunnyvale*, from the Navy Pacific Missile Range Facility, Hawaii. Deck crews of Air Force personnel were trained to operate the ships along with qualified squadron CH-3B helicopter pilots.

Apr 21, 1965—The *Longview* sailed for the first time under Air Force control under the tactical command of squadron Capt Ellsworth Campbell.

May 25, 1965—The *Sunnyvale* sailed for the first time under Air Force control with CH-3B helicopters from the squadron.

Mar 10, 1966—The 6594th Recovery Control Group was redesignated to the 6594th Test Group.

Jun 1966—The third JC-130H aircraft was received by the squadron (that already had eleven JC-130Bs) bringing the squadron's total JC-130 fleet to fourteen aircraft.

Jul 21, 1966—Two JC-130s participated in the mid-Pacific recovery force during the NASA Gemini 10 mission.

Sep 15, 1966—Two JC-130s participated in the mid-Pacific recovery force during the NASA Gemini 11 mission.

Nov 12, 1966—Program 241 was redesignated to Program 846.

Nov 15, 1966—Two JC-130s participated in the mid-Pacific recovery force during the NASA Gemini 12 mission.

Jan 26, 1967—Maj Warren Schensted piloted the recovery of RV-2 from Corona 114. This was the final Corona recovery by one of the original C-119 aircraft commanders of the 6593d.

Jan 1967—The squadron flew its final Ashcan recovery mission.

Jun 1967—The squadron accumulated its 5,000th aerial recovery with the JC-130.

Sep 9, 1967—A squadron JC-130 recovered the NASA BioSatellite II capsule.

Jul 7, 1969—Capt Larry Dement piloted a CH-3B helicopter that deployed a pararescue team into the Pacific to recover a NASA Biosatellite capsule.

Second half of 1969—The squadron accumulated its 10,000th aerial recovery.

Jan 25, 1972—A squadron CH-3B helicopter (S/N 62-12572) was forced to make a water landing and eventually sank. The crew evacuated the helicopter and was rescued by the Coast Guard.

May 31, 1972—Capt Donald Hard piloted the recovery of RV-2 from Corona 145. This was the final mission of the Corona Program.

Jul 1, 1972—The 6593d Test Squadron (Special) was inactivated (HQ AFSC Special Order G-82, June 23, 1972). The squadron personnel and its resources were reassigned into the 6594th Test Group.

1964 Mission Statement of the 6593d Test Squadron (Special)

General:

Develop and maintain a capability to effect the aerial recovery of a capsule ejected from an orbiting satellite.

    a.   Provide liaison assistance to the Recovery Control Group.

    b.   Monitor on-site training programs and conduct unit training.

    c.   Report to Recovery Control Group the status of recovery forces and the status of recovery operations.

    d.   Report to Recovery Control Group complete details concerning recovery conditions of capsule.

    e.   Conduct extended surface search and assist in surface recovery as directed by the Recovery Control Group.

    f.   Perform field test and evaluation of recovery equipment.

Squadron Commanders of the 6593d Test Squadron (Special)

Period	Name
August 1, 1958 - July 9, 1962	Maj Joseph G. Nellor
July 9, 1962 - July 1, 1963	Lt Col Owen F. Pratt, Jr.
July 1, 1963 - June 14, 1965	Lt Col Joe B. Thomson, Jr.
June 14, 1965 - July 13, 1966	Lt Col Grover P. Moore, Jr.
July 13, 1966 - July 14, 1968	Lt Col Harold B. Owens
July 14, 1968 - October 9, 1970	Lt Col Paul Stinson
October 9, 1970 - October 31, 1971	Lt Col Marshall H. Fletcher
October 13, 1971 - July 1, 1972	Lt Col Edward T. Lynch, Jr.

1958 Aircrews of the 6593d Test Squadron (Special)

A Flight

C-119J (Tail No. 18037)

Pilot	Capt Mitchell, Harold E.
Copilot	Capt Clawson, Charles B.
Navigator	1st Lt Counts, Robert D.
Crew Engineer	SSgt Hurst, Arthur P.
Winch Operator	TSgt Bannick, Louis F.
Loadmaster	SSgt Harmon, Algaene
Loadmaster	A1C Johnson, Walter
Loadmaster	A1C Gurganious, Billy N.
Loadmaster	A2C Hill, Daniel R.

C-119J (Tail No. 18038)

Pilot	Capt Wilson, Jack R.
Copilot	Capt Grafe, Arthur H.
Navigator	1st Lt Kusunoki, Wilfred H.
Crew Engineer	TSgt Sims, Rayvaugn M.
Winch Operator	TSgt Young, Warren B.
Loadmaster	TSgt Webb, Joseph C.
Loadmaster	A2C Vaugn, Kenneth R.
Loadmaster	A2C Dinwiddie, William H.
Loadmaster	A2C Miller, Glenwood F.

C-119J (Tail No. 18039)

Pilot	Capt Shinnick, Lawrence W.
Copilot	1st Lt Curtin, Donald R.
Navigator	1st Lt Dorton, Bobby R.
Crew Engineer	TSgt Hollifield, Edward L.

C-119J (Tail No. 18039)

Winch Operator	TSgt Sojda, Stanley W.
Loadmaster	SSgt Knight, Haynie G.
Loadmaster	SSgt Turner, Aubrey B.
Loadmaster	A1C Slaton, Forrest G.
Loadmaster	A2C Ponder, Herbert L.

C-119J (Tail No. 18115)

Pilot	Capt McCullough, James P.
Copilot	1st Lt Adams, Jarvis M.
Navigator	1st Lt Linseisen, Frank J.
Crew Engineer	TSgt Powell, Otis L.
Winch Operator	MSgt Ramsey, William S.
Loadmaster	SSgt Bradley, Lawrence G.
Loadmaster	SSgt Williams, Jefferson B.
Loadmaster	A2C James, Frank
Loadmaster	A3C Brown, Donald R.

B Flight

C-119J (Tail No. 18043)

Pilot	Capt Brewton, James A.
Copilot	1st Lt Deere, William J.
Navigator	1st Lt Keck, Charles H.
Crew Engineer	TSgt Hall, Charles M.
Winch Operator	TSgt Cross, James O.
Loadmaster	SSgt Caling Herman C.
Loadmaster	SSgt Mills, Thomas E.
Loadmaster	SSgt Phillips, Thomas E.
Loadmaster	A2C Jackson, Norvell

1958 Aircrews of the 6593d Test Squadron (Special)

C-119J (Tail No. 18045)

Pilot	Capt Hines, Thomas F.
Copilot	Capt Fortune, Vincent J.
Navigator	Capt Upchurch, Lind D.
Crew Engineer	TSgt Brown, Wilbur R.
Winch Operator	SSgt Muehlberger, James P.
Loadmaster	SSgt Culpepper, William B.
Loadmaster	SSgt Gusta, Kenneth, L.
Loadmaster	A2C Gaudio, Ralph
Loadmaster	A3C Hall, Gene G.

C-119J (Tail No. 18049)

Pilot	Capt Mosher, Edward H.
Copilot	Capt Conn, James E.
Navigator	1st Lt Anderson, Everett E.
Crew Engineer	SSgt Beckwith, John E.
Winch Operator	TSgt Stanberry, Willie
Loadmaster	TSgt Cruise, Jack H.
Loadmaster	TSgt Head, Emory M.
Loadmaster	TSgt Kenyon, Francis C.
Loadmaster	SSgt McKain, George D.

C-119J (Tail No. 18050)

Pilot	Capt Mason, Lynwood G.
Copilot	Capt Luber, Howard E.
Navigator	1st Lt Michelini. Walter L.
Crew Engineer	SSgt Glansbeek, Bernard G.
Winch Operator	TSgt Shields, Marvin L.
Loadmaster	SSgt Bryan, Albert J.
Loadmaster	SSgt Stebbins, Fred L.
Loadmaster	A3C Johnson, Owen L.
Loadmaster	A3C Santana, Hector

C-119J (Tail No. 18042)

Pilot	Capt Schensted, Warren C.
Copilot	1st Lt Clifton, Robert B.
Navigator	1st Lt Ludwick, Jack W.
Crew Engineer	TSgt Jenkins, Elbert P.
Winch Operator	TSgt Champion, Leonard F.
Loadmaster	SSgt Anderson, Billy D.
Loadmaster	SSgt Aragon, Matias V.
Loadmaster	A1C Lansberry, John B.
Loadmaster	A2C Dorigan, Charles J.

C-119J (Tail No. 18041)

Pilot	Capt Parker, Jack O.
Copilot	1st Lt Barrow, Floyd P.
Navigator	1st Lt Radel, Andrew A.
Crew Engineer	TSgt Hendon, Billie
Winch Operator	SSgt Stinnett, Hansel D.
Loadmaster	SSgt Klein, Kenneth W.
Loadmaster	SSgt Kosmatka, John F.
Loadmaster	A2C Hackworth, Donald E.
Loadmaster	A2C Stewart, Jim D.

Discoverer/Corona Recovery Pilots of the 6593d

This list only includes the successful Discoverer/Corona missions.

Mission	Date of Recovery	Pilot
Discoverer 13	Aug 11, 1960	Ocean Recovery
Discoverer 14	Aug 19, 1960	Capt Harold Mitchell
Discoverer 17	Nov 14, 1960	Capt Gene Jones
Discoverer 18	Dec 10, 1960	Capt Gene Jones
Discoverer 25	Jun 19, 1960	Ocean Recovery
Discoverer 26	Jul 9, 1961	Capt Jack Wilson
Discoverer 29	Sep 1, 1961	Ocean Recovery
Discoverer 30	Sep 14, 1961	Capt Warren Schensted
Discoverer 32	Oct 14, 1961	Capt Warren Schensted
Discoverer 35	Nov 16, 1961	Capt James McCullough
Discoverer 36	Dec 16, 1961	Ocean Recovery
Discoverer 38	Mar 3, 1962	Capt Jack Wilson
Corona 39	Apr 19, 1962	Maj James Brewton
Corona 41	May 19, 1962	Capt Thomas Hines
Corona 42	Jun 1, 1962	Maj James Brewton
Corona 44	Jun 25, 1962	Maj Gene Jones
Corona 45	Jul 1, 1962	Capt Vernon Betteridge
Corona 46	Jul 22, 1962	Maj Thomas Hines
Corona 47	Jul 31, 1962	Maj James Brewton
Corona 48	Aug 5, 1962	Maj Thomas Hines
Corona 49	Sep 1, 1962	Maj Gene Jones
Corona 51	Sep 18, 1962	Capt Jack Wilson
Corona 52	Oct 2, 1962	Maj Thomas Hines
Corona 53	Oct 13, 1962	Capt James McCullough
Corona 55	Nov 9, 1962	Maj James Brewton
Corona 56	Nov 28, 1962	Capt Stephen Calder
Corona 58	Dec 17, 1962	Capt Walter Milam
Corona 59	Jan 11, 1963	Ocean Recovery

Mission	Date of Recovery	Pilot
Corona 62	Apr 4, 1963	Capt Stephen Calder
Corona 64	May 20, 1963	Ocean Recovery
Corona 65	Jun 14, 1963	Capt Dale Palmer
Corona 66	Jun 29, 1963	Capt Vernon Betteridge
Corona 67	Jul 22, 1963	Capt James Varnadoe
Corona 68	Aug 1, 1963	Capt Walter Milam
Corona 69	Aug 28, 1963	Capt Dale Palmer
Corona 70	Sep 1, 1963	Capt Walter Milam
Corona 71	Sep 26, 1963	Maj Thomas Hines
Corona 72	Nov 2, 1963	Capt Richmond Apaka
Corona 75	Nov 26, 1963	Maj Jack Wilson
Corona 76, RV-1	Feb 19, 1964	Capt Walter Milam
Corona 76, RV-2	Feb 22, 1964	Capt Jeremiah Collins
Corona 79, RV-1	Jun 8, 1964	Maj Jack Wilson
Corona 79, RV-2	Jun 12, 1964	Maj Jack Wilson
Corona 80	Jun 19, 1964	Capt Charles Young
Corona 81, RV-1	Jun 23, 1964	Capt Charles Young
Corona 81, RV-2	Jun 27, 1964	Maj Edwin Bayer
Corona 82, RV-1	Jul 14, 1964	Capt James Varnadoe
Corona 82, RV-2	Jul 18, 1964	Capt Jeremiah Collins
Corona 83, RV-1	Aug 9, 1964	Capt Albert Muller
Corona 83, RV-2	Aug 14, 1964	Capt James McDonald
Corona 84	Aug 27, 1964	Capt Richmond Apaka
Corona 85, RV-1	Sep 19, 1964	Capt Richmond Apaka
Corona 85, RV-2	Sep 24, 1964	Maj Edwin Bayer
Corona 86, RV-1	Oct 9, 1964	Capt James Varnadoe
Corona 87, RV-1	Oct 21, 1964	Capt Jeremiah Collins
Corona 87, RV-2	Oct 23, 1964	Ocean Recovery
Corona 88, RV-1	Nov 6, 1964	Capt James McDonald
Corona 88, RV-2	Nov 7, 1964	Capt Dale Palmer
Corona 89, RV-1	Nov 22, 1964	Capt Dale Palmer
Corona 89, RV-2	Nov 27, 1964	Maj Edwin Bayer
Corona 90, RV-1	Dec 24, 1964	Capt James Varnadoe

Mission	Date of Recovery	Pilot
Corona 90, RV-2	Dec 30, 1964	Capt James Varnadoe
Corona 91, RV-1	Jan 20, 1964	Capt Jeremiah Collins
Corona 91, RV-2	Jan 25, 1965	Capt Albert Muller
Corona 92, RV-1	Mar 2, 1965	Capt Douglas Sliger
Corona 92, RV-2	Mar 6, 1965	Capt James McDonald
Corona 93, RV-1	Mar 29, 1965	Maj Edwin Bayer
Corona 93, RV-2	Apr 1, 1965	Capt Douglas Sliger
Corona 94, RV-1	May 4, 1965	Capt James Varnadoe
Corona 95, RV-1	May 23, 1965	Maj Edwin Bayer
Corona 95, RV-2	May 28, 1965	Capt Albert Muller
Corona 96, RV-1	Jun 15, 1965	Capt William Vipraio
Corona 96, RV-2	Jun 16, 1965	Capt Albert Muller
Corona 97, RV-1	Jul 24, 1965	Capt Don Olsen
Corona 97, RV-2	Jul 29, 1965	Capt Douglas Sliger
Corona 98, RV-1	Aug 22, 1965	Capt Albert Muller
Corona 98, RV-2	Aug 26, 1965	Capt William Vipraio
Corona 100, RV-1	Sep 27, 1965	Maj Dale Palmer
Corona 100, RV-2	Oct 2, 1965	Capt James McDonald
Corona 101, RV-1	Oct 10, 1965	Maj Edwin Bayer
Corona 101, RV-2	Oct 15, 1965	Capt Nicola Ruscetta
Corona 102, RV-1	Nov 2, 1965	Capt Don Olsen
Corona 102, RV-2	Nov 7, 1965	Capt Douglas Sliger
Corona 103, RV-1	Dec 10, 1965	Capt Herbert Bronson
Corona 103, RV-2	Dec 11, 1965	Capt William Vipraio
Corona 104, RV-1	Dec 29, 1965	Capt Albert Muller
Corona 104, RV-2	Jan 1, 1966	Maj Dale Palmer
Corona 105, RV-1	Feb 7, 1966	Capt Nicola Ruscetta
Corona 105, RV-2	Feb 12, 1966	Maj Don Olsen
Corona 106, RV-1	Mar 15, 1966	Maj Walter Milam
Corona 106, RV-2	Mar 19, 1966	Maj Albert Muller
Corona 107, RV-1	Apr 14, 1966	Maj Walter Milam
Corona 107, RV-2	Apr 18, 1966	Capt John Cahoon
Corona 109, RV-1	May 29, 1966	Maj Edwin Bayer

Mission	Date of Recovery	Pilot
Corona 109, RV-2	Jun 4, 1966	Capt Nicola Ruscetta
Corona 110, RV-1	Jun 26, 1966	Capt Joseph Modicut
Corona 110, RV-2	Jul 1, 1966	Capt William Vipraio
Corona 111, RV-1	Aug 16, 1966	Capt Jack Wenning
Corona 111, RV-2	Aug 23, 1966	Capt John Cahoon
Corona 112, RV-1	Sep 25, 1966	Capt Richard Schofield
Corona 112, RV-2	Sep 30, 1966	Capt Herbert Bronson
Corona 113, RV-1	Nov 12, 1966	Capt William Vipraio
Corona 113, RV-2	Nov 20, 1966	Capt Jack Parker
Corona 114, RV-1	Jan 19, 1967	Maj Jack Wenning
Corona 114, RV-2	Jan 26, 1967	Maj Warren Schensted
Corona 115, RV-1	Feb 28, 1967	Capt Richard McDevitt
Corona 115, RV-2	Mar 6, 1967	Capt John Cahoon
Corona 116, RV-1	Apr 4, 1967	Maj Harlan Gurney
Corona 116, RV-2	Apr 8, 1967	Capt Herbert Bronson
Corona 117, RV-1	May 15, 1967	Capt Richard Schofield
Corona 117, RV-2	May 24, 1967	Capt Joseph Modicut
Corona 118, RV-1	Jun 22, 1967	Maj Jack Wenning
Corona 118, RV-2	Jul 1, 1967	Ocean Recovery
Corona 119, RV-1	Aug 15, 1967	Capt Robert Larison
Corona 119, RV-2	Aug 22, 1967	Capt Edgar Pressgrove
Corona 120, RV-1	Sep 21, 1967	Maj Lester McChristian
Corona 120, RV-2	Sep 28, 1967	Maj Nocholas Ratiani
Corona 121, RV-1	Nov 9, 1967	Capt Edgar Pressgrove
Corona 121, RV-2	Nov 12, 1967	Maj Harlan Gurney
Corona 122, RV-1	Dec 14, 1967	Capt Richard Schofield
Corona 122, RV-2	Dec 22, 1967	Maj Paul Martin
Corona 123, RV-1	Jan 31, 1968	Maj Kenneth Gilbert
Corona 123, RV-2	Feb 7, 1968	Capt Edgar Pressgrove
Corona 124, RV-1	Mar 22, 1968	Capt Joseph Modicut
Corona 124, RV-2	Mar 29, 1968	Capt Albert Kaiser
Corona 125, RV-1	May 7, 1968	Maj Robert Miller

Discoverer/Corona Recovery Pilots of the 6593d

Mission	Date of Recovery	Pilot
Corona 125, RV-2	May 15, 1968	Maj Nocholas Ratiani
Corona 126, RV-1	Jun 29, 1968	Maj Harlan Gurney
Corona 126, RV-2	Jul 5, 1968	Maj Ralph Gauthier
Corona 127, RV-1	Aug 14, 1968	Capt Richard Schofield
Corona 127, RV-2	Aug 22, 1968	Maj Paul Martin
Corona 128, RV-1	Sep 27, 1968	Capt Albert Kaiser
Corona 128, RV-2	Oct 2, 1968	Maj Nocholas Ratiani
Corona 129, RV-1	Nov 12, 1968	Capt Joseph Modicut
Corona 129, RV-2	Nov 21, 1968	Maj Ralph Gauthier
Corona 130, RV-1	Dec 18, 1968	Maj Lester McChristian
Corona 130, RV-2	Dec 23, 1968	Capt Edgar Pressgrove
Corona 131, RV-1	Feb 1, 1969	Capt Richard Schofield
Corona 131, RV-2	Feb 14, 1969	Capt Robert Brenci
Corona 132, RV-1	Mar 21, 1969	Maj Ralph Gauthier
Corona 132, RV-2	Mar 22, 1969	Maj Miller Peeler
Corona 133, RV-1	May 9, 1969	Maj Robert Miller
Corona 133, RV-2	May 18, 1969	Maj Paul Martin
Corona 134, RV-1	Aug 2, 1969	Ocean Recovery
Corona 134, RV-2	Aug 12, 1969	Maj Robert Thornquist
Corona 135, RV-1	Sep 29, 1969	Maj Edward Lynch
Corona 135, RV-2	Oct 7, 1969	Maj Lester McChristian
Corona 136, RV-1	Dec 11, 1969	Capt Edgar Pressgrove
Corona 136, RV-2	Dec 21, 1969	Maj Richard Bussey
Corona 137, RV-1	Mar 12, 1970	Maj Miller Peeler
Corona 137, RV-2	Mar 23, 1970	Maj Robert Thornquist
Corona 138, RV-1	May 31, 1970	Capt Robert Brenci
Corona 138, RV-2	June 8, 1970	Capt Bobbie Mitchell
Corona 139, RV-1	Jul 30, 1970	Capt Marshall Eto
Corona 139, RV-2	Aug 10, 1970	Lt Col William Scott
Corona 140, RV-1	Nov 27, 1970	Capt Thomas Rauk
Corona 140, RV-2	Dec 7, 1970	Maj John Swatek
Corona 142, RV-1	Mar 31, 1971	Maj Maurice Alford
Corona 142, RV-2	Apr 9, 1971	Maj Miller Peeler

Mission	Date of Recovery	Pilot
Corona 143, RV-1	Sep 18, 1971	Maj Robert Jefferies
Corona 143, RV-2	Sep 29, 1971	Capt Mike Hollomon
Corona 144, RV-1	May 1, 1972	Lt Col James McDonald
Corona 144, RV-2	May 8, 1972	Maj Harry Boyd
Corona 145, RV-1	May 27, 1972	Capt Thomas Rauk
Corona 145, RV-2	May 31, 1972	Capt Donald Hard

Original Assigned Aircraft of the 6593d Test Squadron (Special)

Aircraft	Serial Number	Delivery Date
C-119J	51-8039	Aug 21, 1958
C-119J	51-8043	Aug 29, 1958
C-119J	51-8042	Aug 29, 1958
C-119J	51-8038	Sep 4, 1958
C-119J	51-8050	Sep 8, 1958
C-119J	51-8049	Sep 11, 1958
C-119J	51-8045	Sep 18, 1958
C-119J	51-8037	Sep 19, 1958
C-119J	51-8115	Sep 25, 1958
C-119J	51-8041*	Unknown

*C-119J flown by OL-1 of the 6594th Aerospace Test Wing at Edwards AFB

Fairchild Aircraft produced 1,100 Flying Boxcar C-119 cargo airplanes between 1947 and 1955. The C-119 was flown in the Korean and Vietnam Wars.

Specifications:

Span:	109 feet 3.25 inches
Length:	86 feet 5.75 inches
Height:	26 feet 7.75 inches
Weight:	66,900 pounds maximum
Engines:	Two Wright R-3350s of 3,500 horsepower
Cost:	$590,000

Performance:

Maximum speed:	290 miles per hour/252 knots
Cruising speed:	200 miles per hour/174 knots
Range:	2000 statute miles/1738 nautical miles
Service ceiling:	30,000 feet

Flying Hours of the 6593d Test Squadron (Special)

Period	Aircraft	Hours
Sep-Dec 1958	C-119J	1172
Jan-Jun 1959	C-119J	1431
Jul-Dec 1959	Not available	
1960	Not available	
Jan-Jun 1961	C-119	1722
	JC-130B	365
Jul-Dec 1961	JC-130B	1362
	C-119J	949
	CH-21B	112
Jan-Jun 1962	JC-130B	2127
	CH-21B	412
Jul-Dec 1962	JC-130B	2583
Jan-Jun 1963	JC-130B	2056
	CH-21B	447
Jul-Dec 1963	JC-130B	2230
	CH-21B	375
	CH-3B	162
Jan-Jun 1964	JC-130B	2914
	CH-3B	433
Jul-Dec 1964	JC-130B	2426
	CH-3B	397
Jan-Jun 1965	JC-130B	2583
	CH-3B	500

Period	Aircraft	Hours
Jul-Dec 1965	JC-130B	2919
	JC-130H	611
	CH-3B	1082
Jan-Jun 1966	JC-130B	2816
	JC-130H	986
	CH-3B	1082
Jul-Dec 1966	JC-130B	3081
	JC-130H	958
	CH-3B	854
Jan-Jun 1967	JC-130B	2887
	JC-130H	1047
	CH-3B	939
Jul-Dec 1967	JC-130B	2869
	JC-130H	1160
	CH-3B	954
Jan-Jun 1968	JC-130B	3039
	JC-130H	1065
	CH-3B	995
Jul-Dec 1968	JC-130B	2943
	JC-130H	844
	CH-3B	971
Jan-Jun 1969	JC-130B	2670
	JC-130H	940
	CH-3B	957
Jul-Dec 1969	JC-130B	2546
	JC-130H	1013
	CH-3B	910

Flying Hours 6593d Test Squadron (Special)

Period	Aircraft	Hours
Jan-Jun 1970	JC-130B	2758
	JC-130H	857
	CH-3B	884
Jul-Dec 1970	JC-130B	2658
	JC-130H	964
	CH-3B	881
Jan-Jun 1971	JC-130B	2584
	JC-130H	1103
	CH-3B	922
Jul-Dec 1971	JC-130B	2896
	JC-130H	768
	CH-3B	917
Jan-Jun 1972	JC-130B	2842
	JC-130H	808
	CH-3B	806

Awards and Honors of the 6593d Test Squadron (Special)

Period	Awards
1960	MacKay Trophy
Aug 1958-Apr 1962	Air Force Outstanding Unit Award

CITATION TO ACCOMPANY THE AWARD OF

THE AIR FORCE OUTSTANDING UNIT AWARD

TO

6593D TEST SQUADRON (SPECIAL)

The 6593d Test Squadron (Special), Air Force Systems Command, distinguished itself by exceptionally meritorious service from 1 August 1958 to 30 April 1962. During this period, through the development and application of techniques for the recovery and return of space capsules ejected from orbiting satellites, the 6593d Test Squadron (Special) has materially aided in enhancing the prestige of the United States during a time when success in this field was of great urgency. The initiative, resourcefulness and selfless devotion to duty displayed by the personnel of the 6593d Test Squadron (Special) reflect great credit upon themselves and the Unites States Air Force.

Squadron Emblem

The Air Force approved the 6593d Test Squadron (Special) emblem on July 14, 1961

Description:

"On a white rectangle bordered red, an Air Force golden yellow lightning flash bendwise throughout, pointing upward, edged red, surmounted by an Air Force blue falcon in flight, beak and talons Air Force golden yellow, outlines Air Force blue, details and markings white, his talons grasping a red parachute dropping to base."

Significance:

"The emblem is symbolic of the squadron and its mission. Against a lightning streak representing the speed and accuracy of the unit's test operations, a falcon (representing the sharp mind and keen sight required by crew members, and the flying skill exhibited by the pilots) is displayed in flight and grasping a parachute (symbolizing the ability of the aircraft to retrieve a parachuting satellite from celestial regions). Because the falcon is known for its fierce determination to follow through on anything it undertakes, it represents the spirit of the squadron to succeed at the most unusual aerial recovery in modern aviation history. The emblem bears the Air Force colors, ultramarine blue and golden yellow, and the national colors, red white, and blue."

Documents

After Captain Mitchell's successful air retrieval of the Discoverer 14 capsule, he and his crew were honored with several awards. The documents that follow are the letters acknowledging their extraordinary achievements.

The general orders that activated the 6593d Test Squadron (Special)

HEADQUARTERS
AIR RESEARCH AND DEVELOPMENT COMMAND
United States Air Force
Andrews Air Force Base
Washington 25, D. C.

GENERAL ORDERS)
NUMBER   38)

22 July 1958

SECTION

DESIGNATION, ORGANIZATION AND ASSIGNMENT OF USAF UNIT. I

I. DESIGNATION, ORGANIZATION AND ASSIGNMENT OF USAF UNIT.

1. Effective 1 August 1958, the 6593d Test Squadron (Special) is designated and organized at Hickam Air Force Base, Territory of Hawaii with assignment to Air Research and Development Command. The 6593th Test Squadron (Special) is attached to Detachment Number 2, Headquarters Air Research and Development Command (Air Force Ballistic Missile Division), for administrative and operational control, and to the 6486th Air Base Wing for administrative and logistic support.

2. Personnel will be furnished by the Commander, Detachment Number 2, Headquarters, Air Research and Development Command (Air Force Ballistic Missile Division) as prescribed in Department of Air Force Message (S) AFOOP-OC-R 53042, 9 July 1958.

3. Equipment is authorized, effective 1 August 1958, under Unit Authorization List Number 6593 6885A.

4. Authority: Department of the Air Force Message (S) AFOOP-OC-R 53042, 9 July 1958, as amended, and Air Force Regulation 20-27, 15 September 1955.

FOR THE COMMANDER:

J. W. SESSUMS, JR.
Major General, USAF
Vice Commander

W. J. ATKINS
Colonel, USAF
Director Administrative Services

A CERTIFIED TRUE COPY:

EUGENE W. DOMBROSKI
Captain, USAF
Administrative Officer

DISTRIBUTION:
RDAO   (3)
AFBMD   (20)
Dir Admin Serv, The Pentagon, Wash 25, D. C. (20)
"A" DISTRIBUTION

Citation, awarding the Distinguished Flying Cross

CITATION TO ACCOMPANY THE AWARD OF

THE DISTINGUISHED FLYING CROSS

TO

HAROLD E. MITCHELL

    Captain Harold E. Mitchell distinguished himself by extraordinary achievement while participating in aerial flight on 19 August 1960, approximately 400 miles Southwest of Hawaii. On that date, Captain Mitchell was the pilot of the C-119 Aircraft which completed the first aerial recovery of a space capsule ejected from an orbiting satellite in the DISCOVERER XIV operation. The outstanding professional ability displayed by Captain Mitchell in the successful accomplishment of this special mission of international significance reflects great credit upon himself and the United States Air Force.

Official orders, awarding Mitchell and crew the Distinguished Flying Cross and the Air Medal, respectively

```
        General Orders 61, DAF, is the last of the series for 1960

                    DEPARTMENT OF THE AIR FORCE
                            WASHINGTON

SPECIAL ORDERS)                                          12 January 1961
NUMBER    G-1)

1. DP, GEN NATHAN F TWINING, 10A, is awarded the Distinguished Service
Medal (Second Oak Leaf Cluster) for exceptionally meritorious and
distinguished service in a position of great responsibility from 15 Aug 57
to 30 Sep 60.

2. DP, COL RIVERA ARCOS, Uruguayan AF, is awarded the Legion of Merit
(Degree of Officer) for exceptionally meritorious conduct in the performance
of outstanding service from Nov 57 to Jan 60.

3. DP, CAPT HAROLD E MITCHELL, 39402A, is awarded the Distinguished Flying
Cross for extraordinary achievement while participating in aerial flight on
19 Aug 60.

4. DP, each of the following is awarded the Airman's Medal for heroism
involving voluntary risk of life under conditions not involving actual
conflict with an armed enemy during period indicated:

              CWO, W2 EARL R SLATTERY, W2205506, USA
                       12 Jul 60 to 16 Jul 60
              MSGT EDWIN H EATON, RA31449412, USA
                       12 Jul 60 to 16 Jul 60
              TSGT JAMES J FAYED, AF13166237
                            20 Jun 60
              TSGT JAMES E HOLCOMB, AF14322003
                            20 Jun 60
              SSGT GORDON M CLARE, AF12427936
                            14 Aug 60
              A1C JAMES J PENNINGTON, AF13429914
                            11 Sep 60

5. DP, each of the following is awarded the Air Medal for meritorious
achievement while participating in aerial flight on 19 Aug 60:

              CAPT RICHMOND A APAKA, AO3009748
              1STLT ROBERT W COUNTS, AO3065536
              TSGT LOUIS F BANNICK, AF17021127
              SSGT ALGAENE HARMON, AF14363461
              SSGT ARTHUR P HURST, AF34730693
              A1C GEORGE W DONAHOU, AF54048386
              A2C LESTER L BEALE, JR, AF11337082
              A2C DANIEL R HILL, AF13606761

                                                              G-1
```

Announcement, Aerospace Primus Club inductees

```
SO G-1, DAF, 12 Jan 61, cont:

BY ORDER OF THE SECRETARY OF THE AIR FORCE:
```

```
                                            THOMAS D. WHITE
                                            Chief of Staff


J. L. TARR                                  DISTRIBUTION
Colonel, USAF                                    GO
Director of Administrative Services
```

NEWS RELEASE

UNITED STATES AIR FORCE
AIR FORCE SYSTEMS COMMAND
Office Of Information
Andrews Air Force Base
Washington, D.C. 20331
Area Code 301
981 Ext. 4137 - 4138

RELEASE NO.

FOR RELEASE:

AIR FORCE SYSTEMS COMMAND
AEROSPACE PRIMUS CLUB

The most exclusive club on earth, the Aerospace Primus Club was established on Aug. 26, 1960 by General B. A. Schriever, commander of Air Force Systems Command. In the four years of its existence only seven new members have been added to the 13 charter members. Major Robert M. White became the first member to be honored twice for his achievements.

Objectives of the club -- stated in its charter -- are to promote original accomplishment in the discovery, development, testing and use of techniques and equipment in furtherance of the aerospace program and to honor the first individuals to attain or participate in aerospace accomplishments deemed to be of historical significance.

During Aug. 12-19, 1960 charter members were recognized for three spectacular aerospace accomplishments. These were: piloting of the X-15 to a record altitude of 136,500 feet; the first aerial recovery of a space capsule from an orbiting satellite; and the ascent of an open gondola balloon to an altitude of 102,800 feet coupled with the longest free-fall in aviation history and the highest parachute jump on record.

Membership is restricted to military and civilian members of the U. S. Air Force who have accomplished significant aerospace firsts. A special Primus Club board, made up of top military and civilian officials at Headquarters Air Force Systems Command, meets as required to consider nominations for new members. All actions of the Board are submitted to the Air Force Systems Command Commander for his personal approval.

FORGING MILITARY SPACEPOWER

Colonel Alvan N. Moore

For directing a significant aerospace historical first on 19 August 1960. As a result of Colonel Moore's superb planning and leadership, a C-119 aircraft succeeded in accomplishing the first aerial recovery of a space capsule ejected from an orbiting satellite in the Discoverer XIV Operation.

Captain Harold E. Mitchell

For accomplishing a significant aerospace historical first on 19 August 1960 approximately 400 miles southwest of Hawaii. On that date, Captain Mitchell was the pilot of the C-119 aircraft which completed the first aerial recovery of a space capsule from an orbiting satellite. Sighting the capsule parachute at an altitude of 16,000 feet, Captain Mitchell maneuvered his aircraft into an ideal position for the aerial catch and on the third attempt succeeded in recovering the capsule at an altitude of 8500 feet.

Staff Sergeant Wendell King

For accomplishing a significant aerospace historical first on 19 August 1960 approximately 400 miles Southwest of Hawaii. On that date, Sergeant King was Photographer of the C-119 aircraft which completed the first aerial recovery of a space capsule from an orbiting satellite. Sergeant King displayed exceptional skill and professional ability in photographing the successful recovery of the capsule.

Captain Richmond A. Apaka

For accomplishing a significant aerospace historical first on 19 August 1960 approximately 400 miles Southwest of Hawaii. On that date, Captain Apaka was Copilot of the C-119 aircraft which completed the first aerial recovery of a space capsule from an orbiting satellite. Captain Apaka displayed exceptional skill, professional ability and teamwork in assisting in the successful recovery of the capsule on the third attempt at an altitude of 8500 feet.

First Lieutenant Robert G. Counts

For accomplishing a significant aerospace historical first on 19 August 1960 approximately 400 miles southwest of Hawaii. On that date, Lieutenant Counts was Navigator of the C-119 aircraft which completed the first aerial recovery of a space capsule from an orbiting satellite. Lieutenant Counts displayed exceptional skill, professional ability and teamwork in assisting in the successful recovery of the capsule on the third attempt at an altitude of 8500 feet.

Technical Sergeant Louis F. Bannick

For accomplishing a significant aerospace historical first on 19 August 1960 approximately 400 miles Southwest of Hawaii. On that date, Sergeant Bannick was Winch Operator and Loadmaster of the C-119 aircraft which completed the first aerial recovery of a space capsule from an orbiting satellite. Sergeant Bannick displayed exceptional skill, professional ability and teamwork in assisting in the successful recovery of the capsule on the third attempt at an altitude of 8500 feet.

Staff Sergeant Arthur P. Hurst

For accomplishing a significant aerospace historical first on 19 August 1960 approximately 400 miles Southwest of Hawaii. On that date, Sergeant Hurst was Flight Engineer and Aircraft Maintenance Technician of the C-119 aircraft which completed the first aerial recovery of a space capsule from an orbiting satellite. Sergeant Hurst displayed exceptional skill, professional ability and teamwork in assisting in the successful recovery of the capsule on the third attempt at an altitude of 8500 feet.

Staff Sergeant Algaene Harmon

For accomplishing a significant aerospace historical first on 19 August 1960 approximately 400 miles Southwest of Hawaii. On that date, Sergeant Harmon was Loadmaster of the C-119 aircraft which completed the first aerial recovery of a space capsule from an orbiting satellite. Sergeant Harmon displayed exceptional skill, professional ability and teamwork in assisting in the successful recovery of the capsule on the third attempt at an altitude of 8500 feet.

Airman First Class George W. Donahou

For accomplishing a significant aerospace historical first on 19 August 1960 approximately 400 miles Southwest of Hawaii. On that date, Airman Donahou was Loadmaster of the C-119 aircraft which completed the first aerial recovery of a space capsule from an orbiting satellite. Airman Donahou displayed exceptional skill, professional ability and teamwork in assisting in the successful recovery of the capsule on the third attempt at an altitude of 8500 feet.

Airman Second Class Lester L. Beale

For accomplishing a significant aerospace historical first on 19 August 1960 approximately 400 miles Southwest of Hawaii. On that date, Airman Beale was Loadmaster of the C-119 aircraft which completed the first aerial recovery of a space capsule from an orbiting satellite. Airman Beale displayed exceptional skill, professional ability and teamwork in assisting in the successful recovery of the capsule on the third attempt at an altitude of 8500 feet.

Airman Second Class Daniel R. Hill

For accomplishing a significant aerospace historical first on 19 August 1960 approximately 400 miles Southwest of Hawaii. On that date, Airman Hill was Loadmaster of the C-119 aircraft which completed the first aerial recovery of a space capsule from an orbiting satellite. Airman Hill displayed exceptional skill, professional ability and teamwork in assisting in the successful recovery of the capsule on the third attempt at an altitude of 8500 feet.

Major Robert M. White

For accomplishing a significant aerospace historical first on 12 August 1960 at Edwards Air Force Base, California. On that date, Major White piloted the X-15 aircraft to a record altitude of 136,500 feet. Major White displayed outstanding courage and professional skill in piloting this aircraft to higher altitudes and faster speeds than man had ever obtained.

Captain Joseph W. Kittinger, Jr.

For accomplishing a significant aerospace historical first on 16 August 1960 near Holloman Air Force Base, New Mexico. On that date, Captain Kittinger made an open gondola balloon ascent to the record altitude of 102,000 feet, more than 19 miles above the New Mexico desert. At this altitude, he egressed for a parachute descent by means of an experimental stabilization parachute system free falling for four minutes and thirty-eight seconds to an altitude of 17,500 feet where a deployment of the man recovery parachute occurred.

Technical Sergeant Lonoteo M. Vigare

For accomplishing a significant aerospace historical first on 18 June 1961. On that date Sergeant Vigare, a member of the first Pararescue Scuba Team to recover a space capsule containing valuable scientific equipment, parachuted into the Pacific Ocean northeast of Hawaii, spent the night in the open sea with minimum survival gear and succeeded in recovering the nose cone from the Discoverer XXV Satellite returning from orbital flight.

Staff Sergeant William V. Vargas

For accomplishing a significant aerospace historical first on 18 June 1961. On that date Sergeant Vargas, a member of the first Pararescue Scuba Team to recover a space capsule containing valuable scientific equipment, parachuted into the Pacific Ocean northeast of Hawaii, spent the night in the open sea with minimum survival gear and succeeded in recovering the nose cone from the Discoverer XXV Satellite returning from orbital flight.

Staff Sergeant Ray E. McClure

For accomplishing a significant aerospace historical first on 18 June 1961. On that date Sergeant McClure, a member of the first Pararescue Scuba Team to recover a space capsule containing valuable scientific equipment parachuted into the Pacific Ocean northeast of Hawaii, spent the night in the open sea with minimum survival gear and succeeded in recovering the nose cone from the Discoverer XXV Satellite returning from orbital flight.

Letter, Awarding the 6593d Test Squadron (Special) the MacKay Trophy for 1960

MISSION: The mission of the Squadron is to develop training recovery techniques, and employ these procedures in aerial or surface recovery of scientific components of appropriate re-entry vehicles.

AWARDS: The 6593rd Test Squadron (Special) was recipient of the MacKay Trophy for 1960. Major Joseph G. Nellor, Squadron Commander, in a ceremony held in Washington, D.C., received the trophy and citation on behalf of the 6593rd Test Squadron (Special) from the Secretary of the Air Force, the Honorable Eugene M. Zuckert.[1]

The Citation for the award of the MacKay Trophy is as follows:

"THE MACKAY TROPHY for 1960 is awarded to the 6593rd Test Squadron (Special) for the most meritorious flight of the year. On 19 August 1960, this unit located and recovered the capsule of DISCOVERER XIV, which marked the first time that any nation had recovered an object from orbit while it was still descending, using aerial recovery techniques. This feat was successfully repeated by aerial recovery of the capsules of DISCOVERER XVII on 14 November 1960 and DISCOVERER XVIII on 10 December 1960. The exceptional unit effort and operational effectiveness which made these achievements possible are in the highest traditions of the United States Air Force.

s/t/THOMAS D. WHITE
Chief of Staff

1. See supporting document #1

Letter, 6593d honored with the Air Force Outstanding Unit award

HEADQUARTERS
AIR FORCE SYSTEMS COMMAND
UNITED STATES AIR FORCE
ANDREWS AIR FORCE BASE
WASHINGTON 25, D. C.

4 DEC

REPLY TO
ATTN OF: SCP

SUBJECT: Air Force Outstanding Unit Award - 6593d Test Squadron (Special)

TO: SSD (SSG)
AF Unit Post Office
Los Angeles 45, Calif

1. It gives me great pleasure to forward the elements of the Air Force Outstanding Unit Award for presentation to the personnel of the 6593d Test Squadron (Special). I desire that the presentation be made under conditions which will afford all personnel of the Squadron, their families and friends, an opportunity to attend the ceremony.

2. The members of the 6593d Test Squadron may be justifiably proud of this recognition, since very few organizations in non-combat commands have been honored with this award during the past three years. The award signifies a truly outstanding accomplishment, unmatched by any organization of similar composition and mission.

3. I commend all personnel of the 6593d Test Squadron (Special) for their outstanding record and extend best wishes for continued success in the performance of their mission.

B. A. SCHRIEVER
General, USAF
Commander

2 Atch
1. Certificate
2. Citation

Attachment: letter from Major General Ben I. Funk

Ltr, Hq AFSC (SCP), 4 Dec 62, Air Force Outstanding Unit Award - 6593d Test Squadron (Special)

1st Ind (SSC)   22 JAN 1963

Hq SSD, AF Unit Post Office, Los Angeles 45, Calif

TO: 6594th Aerospace Test Wing, Sunnyvale, Calif

1. The exceptional achievement of the 6593d Test Squadron (Special) is highly gratifying to me as Commander, Space Systems Division. In forwarding the elements of the Air Force Outstanding Unit Award for presentation to the personnel of the 6593d Test Squadron, I add my sincere congratulations to those expressed by General Schriever.

2. The unparalleled accomplishments of the 6593d Test Squadron not only bring great honor to each individual in that unit, but reflect great credit upon the Air Force Systems Command and the United States Air Force.

BEN I. FUNK   2 Atch
Major General, USAF   n/c
Commander

Attachment: letter from Colonel W.K. Kincaid

Ltr, Hq AFSC (SCP), 4 Dec 62, Air Force Outstanding Unit Award - 6593 Test Squadron (Special)

2d Ind (TWG) 24 JAN 1963

Hq 6594 Aerospace Test Wing, Sunnyvale, Calif

TO: 6594 Recovery Control Group, APO 953, San Francisco, Calif

The distinguished record of the men of the 6593 Test Squadron (Special) is in consonance with the finest traditions of the military service. Their achievements herald a standard of excellence to be emulated by all who follow the profession of arms.

W. K. KINCAID
Colonel, USAF
Commander

2 Atch
n/c

3   6a

Attachment: letter from Colonel W R Morton

Ltr, Hq AFSC (SCP), 4 Dec 62, Air Force Outstanding Unit Award - 6593 Test Squadron (Special)

3d Ind (TRG)   15 FEB 1963

Hq 6594th Recovery Control Group

TO: 6593d Test Squadron (Special) (TTG)

1. The award bestowed upon the 6593d Test Squadron (Special) and all the members comprising the unit is a manifestation of the tremendous endeavors and invaluable contributions made in the field of aerospace technology through qualities of job proficiency, tenacity of purpose, and teamwork.

2. The endowment of the Outstanding Unit Award is a distinct and cherished honor and should instill an even greater pride in each individual in belonging to the 6593d Test Squadron (Special). I, along with all other members of the Headquarters 6594th Recovery Control Group, am as happy and proud of your achievement as you are.

W R MORTON
Colonel, USAF
Commander

2 Atch
n/c

Glossary

Acronyms & Abbreviations

ADF	Aerial Direction Finder
AFB	Air Force Base
AFBMD	Air Force Ballistic Missile Division
AFS	Air Force Station
APS	Aerial Port Squadron
ARDC	Air Research and Development Command
AWS	Air Weather Service
CIA	Central Intelligence Agency
CO	Commanding Officer
CONUS	Continental United States
DoD	Department of Defense
ECM	Electronic Countermeasures
EDF	Electronic Direction Finding
ETA	Estimated Time of Arrival
ETPD	Estimated Time of Parachute Deployment
fpm	feet per minute
HQ	Headquarters
ICBM	Intercontinental Ballistic Missile
IRAN	Inspection and Repair as Necessary
JATO	Jet Assisted Take Off
MAC	Military Airlift Command
MATS	Military Air Transport Service
mm	millimeter
NAS	Naval Air Station

NASA	National Aeronautics and Space Administration
NCO	Noncommissioned Officer
NCOIC	Noncommissioned Officer in Charge
OIC	Officer in Charge
OL-1	Operating Location Number 1, a detachment from the 6594th Aerospace Test Wing that conducted aerial recovery flight tests at Edwards AFB.
PACAF	Pacific Air Forces
PIP	Projected Impact Point
RAC	Recovery Aircraft Commander, a mission-capable pilot specifically trained and qualified for aerial recovery.
R&D	Research and Development
RV	Reentry Vehicle
SMC	Space and Missile Systems Center
SOP	Standard Operating Procedure
TAC	Tactical Air Command
TDY	Temporary Duty
USNS	United States Naval Ship

Glossary

Term Definitions

Actuators	The two recovery pole mounts in the aircraft that lowered the 34-foot poles and held them in place during a recovery.
Agena	The spacecraft that hosted and orbited the Corona payload
Aircraft commander	The pilot designated as the commander of an aircraft.
Ashcan	A high altitude air sampling program that studied how particles exchange and eventually end up distributing around the world.
Ballpark	The designated 200 by 60-mile primary recovery zone where a satellite capsule and parachute were expected to descend back to earth. Assigned 6593d aircrews patrolled this zone during a recovery mission.
Discoverer	The unclassified cover name for the Corona program.
Dolly	The large yellow hydraulic boom structure mounted on rails in the cargo section of JC-130B and JC-130H aircraft. During a recovery the dolly could be seen extended outside the rear cargo door. It also helped manage the handling of recovered capsules in the aircraft.
Drag Net	Code name given to the recovery effort for the Moby Dick Project.
Genetrix	Balloon reconnaissance program.
Hooks	The pronged grappling hooks that snared the parachutes during an aerial recovery. The hooks were spliced into the trapeze-like recovery loop and attached to the ends of the two 34-foot poles.
Inverted parachute	When a loop contacted a parachute too low during a recovery, the parachute did not collapse and would be towed behind the airplane like a drag chute. Inversion caused significant drag to the airplane and would often break the winch line or make it impossible for the winch to reel the parachute into the airplane.

Loop	The rope (or cable) and spliced recovery hooks in the aerial recovery trapeze. The loop was between the two 34-foot poles that were lowered beneath the recovery aircraft to snare and recover the parachutes and payloads. One end of the loop was attached to the ends of the two recovery poles, and the other end of the loop was fastened to the winch line by a clevis pin.
Moby Dick	High altitude balloon study.
Outfield	The secondary recovery zone 400 miles downrange of the ballpark. Assigned 6593d aircrews patrolled the outfield during a recovery mission, but were less likely to be in position to recover the satellite capsule.
Package	The capsule attached to a parachute for aerial recovery. A package was also called a "payload" or a "bucket."
Pararescue personnel	Air Force frogmen of the 76th Air Rescue and Recovery Squadron who jumped with rafts from helicopters or airplanes to recover floating Corona capsules that landed in the ocean.
Pelican	The nine assigned 6593d aircrew code names and rotating recovery positions during the early Discoverer recoveries. The Pelican 1 crew would be assigned the most likely position in the ballpark to recover the capsule during a mission descending down to the Pelican 9 crew that would be assigned to patrol the least likely sector in the outfield to recover a capsule.
Pole handlers/riggers	Nicknames for the loadmasters involved in aerial recovery.
Poles	The two 34-foot hollow metal recovery poles (with the loop and hooks between them) that were lowered beneath a recovery aircraft during a midair recovery.
Program 162	The program number designation that replaced the name "Discoverer" for the Corona program on March 3, 1962.
Program 241	The renumbered program designation for Corona that replaced Program 162 on June 12, 1964.

Glossary

Program 846	The renumbered program designation for Corona that replaced Program 241 on November 12, 1966.
Rig	The recovery trapeze that was lowered beneath an aircraft to recover parachutes. The rig included the two 34-foot recovery poles with the attached loops and hooks.
Rigged	When the recovery equipment was extended from the aircraft and ready for a recovery.
System	The capsule/payload and attached parachute that was recovered by the 6593d in midair.
Tear through	When an attempted aerial recovery made physical contact with the parachute but failed to make the recovery.
Thor	The launch vehicle that transported Discoverer/Corona into orbit.

Reference Notes

Preface

1. Robert McDonald, *Corona, Between the Sun & the Earth: The First NRO Reconnaissance Eye In Space*, (Bethesda: American Society for Photogrammetry and Remote Sensing, 1997).

2. Curtis Peebles, *The Corona Project*, (Annapolis: Naval Institute Press, 1997).

Chronology of the 6593d Test Squadron (Special)

1. United States Air Force, *The 6594th Test Group, A Chronology 1958-Present*, 1983, pp. 1-6.

2. The dates of the Discoverer/Corona recoveries and the pilots who recovered them were referenced from: Curtis Peebles, *The Corona Project*, (Annapolis, Naval Institute Press, 1997) pp. 272-315.

1964 Mission Statement of the 6593d Test Squadron (Special)

1. United States Air Force, *History of the 6594th Text Group 1 January-30 June 1964*, p. Attachment #5.

Squadron Commanders of the 6593d Test Squadron (Special)

1. United States Air Force, *History of the 6594th Test Group (Draft)*, circa 1989, p. Appendix IIB.

1958 Aircrews of the 6593d Test Squadron (Special)

1. United States Air Force, *History of the 6593d Test Squadron (Special) (ARDC) 1 August-31 December 1958*, p. Supporting Document #9.

Discoverer/Corona Recovery Pilots of the 6593d

1. The dates of the Discoverer/Corona recoveries and the pilots who recovered them were referenced from: Curtis Peebles, *The Corona Project*, (Annapolis, Naval Institute Press, 1997) pp. 272-315.

Original Assigned Aircraft of the 6593d Test Squadron (Special)

1. United States Air Force, *History of the 6594th Test Group (Draft)*, circa 1989, p. Appendix VIII.

2. The C-119J specifications and performance statistics are referenced from the C-119J display at the National Museum of the US Air Force in 2006.

Flying Hours of the 6593d Test Squadron (Special)

1. United States Air Force, *History of the 6593d Test Squadron (Special) (ARDC) 1 August-31 December 1958*, p. Supporting Document #20.

2. United States Air Force, *History of the 6593d Test Squadron (Special) (ARDC) 1 January-30 June 1959*, p. Supporting Document #14.

3. United States Air Force, *History of the 6593d Test Squadron (Special) 1 January-30 June 1961*, p. 14.

4. United States Air Force, *History of the 6593d Test Squadron (Special) 1 July-31 December 1961*, pp. 10, 17.

5. United States Air Force, *History of the 6593d Test Squadron (Special) 1 January-30 June 1962*, pp. 8, 16.

6. United States Air Force, *History of the 6593d Test Squadron (Special) 1 July-31 December 1962*, p. 6.

7. United States Air Force, *History of the 6593d Test Squadron (Special) 1 January-30 June 1963*, pp. 7, 20.

8. United States Air Force, *History of the 6593d Test Squadron (Special) 1 July-31 December 1963*, p. 7.

9. United States Air Force, *History of the 6593d Test Squadron (Special) 1 January-30 June 1964*, p. 4.

10. United States Air Force, *History of the 6593d Test Squadron (Special) 1 July-31 December 1964*, p. 6.

11. United States Air Force, *History of the 6593d Test Squadron (Special) 1 January-30 June 1965*, p. 5.

12. United States Air Force, *History of the 6593d Test Squadron (Special) 1 July-31 December 1965*, p. 9.

13. United States Air Force, *History of the 6593d Test Squadron (Special) 1 January-30 June 1966*, pp. 6, 10.

14. United States Air Force, *History of the 6593d Test Squadron (Special) 1 July-31 December 1966*, p. 3.

15. United States Air Force, *History of the 6593d Test Squadron (Special) 1 January-30 June 1967*, p. 2.

16. United States Air Force, *History of the 6593d Test Squadron (Special) 1 July-31 December 1967*, p. Attachment 1.

17. United States Air Force, *History of the 6593d Test Squadron (Special) 1 January-30 June 1968*, p. Attachment 1.

18. United States Air Force, *History of the 6593d Test Squadron (Special) 1 July-31 December 1968*, p. 5.

19. United States Air Force, H*istory of the 6593d Test Squadron (Special) 1 January-30 June 1969*, p. 4.

20. United States Air Force, *History of the 6593d Test Squadron (Special) 1 July-31 December 1969*, p. 4.

21. United States Air Force, *History of the 6593d Test Squadron (Special) 1 January-30 June 1970*, p. 10.

22. United States Air Force, *History of the 6593d Test Squadron (Special) 1 July-31 December 1970*, p. 3.

23. United States Air Force, *History of the 6593d Test Squadron (Special) 1 January-30 June 1971*, p. 5.

24. United States Air Force, *History of the 6593d Test Squadron (Special) 1 July-31 December 1971*, p. 6.

25. United States Air Force, *History of the 6593d Test Squadron (Special) 1 January-30 June 1972*, p. 6.

Awards and Honors of the 6593d Test Squadron (Special)

1. United States Air Force, *History of the 6593d Test Squadron (Special) 1 January-30 June 1961*, p. 3.

2. United States Air Force, *History of the 6593d Test Squadron (Special) 1 January-30 June 1963*, pp. 3-7.

Squadron Emblem

1. The emblem patch is courtesy of Charles Dorigan.

2. The emblem description and significance statement is courtesy of the Air Force Historical Research Agency at Maxwell AFB, Alabama.

Index

Symbols

1365th Aerial Photo Squadron *142, 205*

A

Aerospace Primus Club *13, 66, 67, 79, 90, 236*
Agena *9, 36, 37, 39, 40, 46, 47, 49, 54, 73, 95, 96, 116, 118, 205, 249*
Ahola, Teuvo (Gus) *60, 61, 63, 79, 122*
Air Force Ballistic Missile Division (AFBMD) *2, 12, 14, 89, 205*
Air Weather Service *199*
All American Engineering Company *5, 17, 25, 26, 32, 80, 92, 108, 111, 112, 128, 131, 146, 166, 175*
Apaka, Richmond A. *9, 11, 48, 53-56, 58, 59, 61, 65, 70, 74, 218*
Apollo *101, 102*
Arends, Leslie C. *13*
Ashcan *79, 97-101, 105, 165, 171, 178, 179, 180, 199, 200, 202-204, 206, 207, 249*

B

Bannick, Louis F. *3, 4, 9-13, 25, 38, 46, 48, 50, 53, 54-59, 62, 64, 72, 80, 89, 142, 213*
Battle, Lee *13, 14*
Beale, Lester L. *3, 4, 48, 53, 55, 58, 60, 62, 65, 89*
Bell, Richard C. *165*
Brewton, James A. *1, 5, 214, 217*

C

Conway, Harry *17, 32, 131*
Corona *146, 183, 197, 207, 217*
Counts, Robert D. *4, 8, 10, 12, 13, 40, 41, 45, 48, 53-55, 59, 64, 70, 79, 200, 201, 203, 213*
Culpepper, William B. *111, 123, 124, 156, 158, 162, 215*
Curtin, Donald R. *105, 142, 144, 147, 158, 162, 169, 213*

D

Discoverer 13 *7-10, 39-46, 48, 53, 54, 60, 73, 82, 83, 85-87, 89, 91, 106, 118-121, 144, 146, 147, 205, 217*

Discoverer 14 *1, 9-15, 46, 47, 49, 50, 52, 53, 57, 60-66, 68-77, 79, 86- 92, 106, 130, 147, 205, 217, 233*
Discoverer 15 *73, 122, 138, 147*
Donahou, George W. *3, 4, 11, 48, 53, 55, 60, 62, 65, 89*
Dorigan, Charles J. *39, 111, 127, 216*
Drag Net *1-3, 84, 127, 128, 249*

E

Edwards Air Force Base *3, 4, 8, 13-16, 20, 26-29, 32, 33, 48, 53, 56, 60, 66, 76, 79, 81, 94-98, 101-106, 108-116, 118, 122, 123, 125, 127-129, 135, 136, 139-142, 144, 148, 150, 151, 157, 160, 162, 163, 165, 168, 169, 175, 178, 192, 193, 199, 200, 202, 203, 205, 223*
Eielson Air Force Base *98, 171, 179, 180, 202, 203, 206*
Eisenhower, Dwight D. *91, 121, 122, 147*
Eto, Marshall *196, 221*

F

Fairchild Aircraft *15-17, 33, 40, 48, 205, 223*
Fort Campbell *1, 15, 127*

G

Garaway Show, Dave *12*
Genetrix *1, 3, 4, 15-17, 79, 92, 167, 249*
Grand Union *1, 105, 110*
Gurganious, Billy N. *3, 19, 213*
Gurney, Harlan L. *199, 220, 221*

H

Hard, Donald G. *183, 207*
Harmon, Algaene *3, 4, 10, 11, 48, 53, 55, 56, 58-60, 62, 65, 80, 88, 89, 213*
Hauenstein, Charles *109*
Hayes Aircraft *16*
Hickam Air Force Base *6, 9, 32, 33, 35, 37, 38, 43, 44, 46, 51, 58, 60-62, 68, 70, 71, 73, 75, 76, 81, 84, 89, 92, 94, 95, 108, 110, 113, 114, 118, 119, 121, 124, 138, 141, 142, 144, 145, 147, 148, 163, 176, 177, 184, 189, 190, 196, 197, 199, 203, 205, 206*

Hill, Daniel R. *3, 4, 11, 15, 80, 89, 103, 140, 164, 213*
Hines, Thomas F. *1, 4, 140, 215, 217, 218*
Hot Hand *2*
Howard Air Force Base *98, 202, 203*
Hurst, Arthur P. *4, 11, 48, 50, 53, 55, 62, 65, 80, 89, 213*

J

Johnson, Lyndon *146*
Jones, Gene W. *4, 27, 73-75, 113, 217*

K

Kershaw, Theodore *1, 2, 3*
King, Wendell *48, 53, 59, 62, 65, 89*
Kirtland Air Force Base *2*
Kodiak, Alaska *1, 10, 53, 96*
Kodiak Tracking Station *53*
Krump, Donald *197*

L

Lefstad, Roy *12*
Lynch, Edward (Ted) *189, 211, 221*

M

Mark 5 parachute *184, 191, 200*
Mark 8 parachute *99, 184, 190, 191*
Mark 9 parachute *192*
Mason, Lynnwood G. (Lindy) *1, 215*
McCullough, James P. *1, 4, 214, 217*
Messerli, Glen *192*
Military Air Transport Service (MATS) *33, 79, 80, 174, 181*
Mitchell, Harold E. *1, 15, 17, 32, 38, 40, 41, 48, 50, 53-61, 64, 73-75, 77, 79, 80, 85, 86, 88, 89, 147, 205, 213, 217, 233*
Moby Dick *79, 84, 92, 97, 99, 102, 249, 250*
Mosher, Edward H. *1, 110, 111, 119, 121, 122, 142, 144, 146, 147, 154, 156, 162, 163, 215*

O

O'Donnell, Emmett (Rosy) *11, 12, 61, 63, 89*
Ogden Air Material Area *16*

P

Parker, Jack O. *109, 110, 142, 162, 216, 220*
Pope Air Force Base *1-3*

R

Ritland, Osmond J. *2, 12*
Rose, Howard *2*

S

Schensted, Warren C. *1, 129, 130, 141, 142, 206, 207, 216, 217, 220*
Schriever, Bernard A. *2, 13, 45, 64-68, 76, 90, 91, 120, 122*
Scott, William *193, 221*
Sewart Air Force Base *3, 15, 33, 48, 105, 127*
Shinnick, Lawrence W. (Larry) *1, 4, 8, 105-107, 109, 213*
Swatek, Jack *193, 221*

T

Telemetry ships *94-96, 113, 114, 139, 178, 179*
Thor *39, 46, 49, 73, 82, 84, 251*

V

Vandenberg Air Force Base *12, 46, 47, 49, 64, 65, 73, 95, 113, 114, 116, 130, 205*

W

White, Thomas D. *12, 45, 61, 120, 121, 147*
Wilson, Jack R. *1, 4, 5, 117, 125, 205, 206, 213, 217, 218*

X

X-15 *12, 13, 66, 101, 144, 157, 159, 160*

www.ingramcontent.com/pod-product-compliance
Lightning Source LLC
Chambersburg PA
CBHW080534170426
43195CB00016B/2561